Foundations of
Experimental Chemistry

Foundations of
Experimental Chemistry

Jubaraj B. Baruah

Professor

Parikshit Gogoi   **Abhilasha**

Department of Chemistry
Indian Institute of Technology, Guwahati.

PharmaMed Press
An imprint of Pharma Book Syndicate
An unit of BSP Books Pvt., Ltd.
4-4-309/316, Giriraj Lane,
Sultan Bazar, Hyderabad - 500 095.

Published by

PharmaMed Press

An imprint of Pharma Book Syndicate

An unit of BSP Books Pvt., Ltd.

4-4-316, Giriraj Lane, Sultan Bazar, Hyderabad - 500 095.

Phone: 040-23445600, 23445688; Fax: 91+40-23445611

E-mail: info@pharmamedpress.com

www.pharmamedpress.com/pharmamedpress.net

ISBN : 978-93-89974-75-1

Preface

The prime concern of a Chemist is to discover and characterise new molecules and materials and to provide explanation to any new property observed in molecules or molecular assemblies. To learn chemistry, a strong background on experimental techniques and ability to correlate to certain physical properties to structure is essential. The complexity involved in chemistry starts with the availability of pure material to make a reproducible observation to the complexity involved in the structural elucidation. The reproducible experimental data followed by adequate supportive information only can form basis of the chemistry. The extent of understanding depends on the accuracy of the results and precision of each experiment. A general observation may arise from an experimental observation or some phenomenon occurring in nature. Extensive experimentation is needed to formulate a general law to make it widely applicable. Any theoretical model needs to be tested with experimental data to understand the exceptions that may be observed from experimental results. There are large number of books which deals with experimental chemistry with different branches and outlook to use for experimental manual. There is not much effort being put to cover up the deficiency that is required to reduce the left out gaps between the branches while studying them independently.

In the present book we made an attempt to make an unified approach among different branches of chemistry, so that the experiments with clear objectives are presented. Generally, laboratory experiments need to fulfil certain criterions, like amount of time required, relevance to a course, understanding at a level, available materials; experiments should be hazard free, and finally motivating. The basic lacuna while performing experiments is the limited way of projecting the objectives, which makes them less attractive. We sorted out the topics of experiments with motivating objectives.

The book comprises of five chapters and each chapter is written with specific implications. **First chapter** deals with the chemistry around two pairs of small molecules to discuss the diversity and complexity in their chemistry so as to imbibe interest on why and how chemistry can be studied with small molecules. The **Second chapter** is on the principles and experiments that would come across in day-to-day life to establish physical properties associated with them. **Third chapter** narrates some of the experiments along with principles of coordination chemistry so that some useful materials can be synthesised and characterised. The definition, utility of preparation and characterisation of compounds with definite logic different types of coordination complexes are discussed. **Fourth chapter** is on the fundamental principles and examples of organic reactions, some of them with specific applications and to understand fundamentals. **Fifth chapter** is on selective contemporary experiments with principles and elucidate their need with a futuristic approach.

The book will cater to the need of final year Honours in Chemistry, B.Tech (Chemistry and Chemical technology) or the new courses like integrated master degree and master degree in Chemistry, Biochemistry, Materials Science, Medicine, Pharmacy and Polymer. The content is such that it can be useful globally to cater fundamental experimental requirements to have foundation in experimental chemistry at higher level. It is written with a thought that it will generate impetus and take experimental chemistry more close to a reader. The writing of this book originated from an opportunity of working with Prof. J. B. Baruah along with Mr. Parikshit and Mrs. Abhilasha, to implement a laboratory course for M.Sc. chemistry. While designing experiments, we decided to make an effort to put the course materials in the form of a book.

Special thanks from J.B.Baruah to his wife (Helen) and son (Jnanbikash) for their constant support and encouragements. Thanks from Abhilasha to her daughter (Udita) who had spared time for her to work. We are thankful to Indian Institute of Technology, Guwahati, for using some of the facilities. Thanks to Mr.Anil Shah and the team of workers in BS Publications who have put constant effort to publish the book.

- Author

Contents

CHAPTER 1

Experiments with Small Molecules

General Introduction

Chemistry is an experimental subject. It comprises of sub disciplines such as Inorganic, Organic, Physical, and Theoretical chemistry. In none of these branches, chemistry has differences in terms of laws and principles. The demarcations are based on the requirement of understanding the fundamentals with different approaches. Irrespective of branch and emphasis, the prime concern of a chemist is to discover and characterize new molecules and materials and to provide explanation to any new property observed in molecules or molecular assemblies. The molecules may also be in the form of macromolecules or self assembled macromolecules. Basic understanding at molecular level is needed to learn chemistry. The understanding may be either in terms of laws, rules that are derived from extensive experimentation. The complexities involved in chemistry, starts with the availability of pure material/s to make reproducible observations to the structural elucidation. Thus, the other sub-branches like pure and applied chemistry comes into picture. Whatever be the purpose of study, the experimental data followed by adequate supportive information only can form the basis of studying chemistry. The extent of understanding lies on the accuracy of the results and precision of each experiment. Every small molecule has wide ranges of chemistry. The perspective and emphasis of every practitioner, mainly researchers are different. In the following part, we take two pairs of simple molecules, namely (i) water and hydrogen sulphide (ii) ammonia and borane, to show the different approaches to study chemistry with them. Different experimental tools required for understanding their physical properties or reactivity are also discussed.

Chemistry with Water and Hydrogen Sulphide

Structural Aspect

Let us discuss the avenues for studying experimental chemistry of two small molecules, water (H_2O) and hydrogen sulphide (H_2S). We can start by saying H₂O and H₂S as two triatomic bent molecules and provide reasons for their bent structures. The Lewis dot structures for both the compounds are similar and both of them have two pairs of non-bonded electrons. Structures can be explained with the aid of valence shell electron pair repulsion theory. They are AB_2E_2 type molecules (A is the central atom, B is atom covalently linked to atom A, and E is lone pair of electrons); adopt non-linear structures (Fig. 1.1). Alternately, one may also say that the elements, oxygen, and sulphur belong to same group in periodic table. Both of them have two hydrogen atoms attached to the oxygen or sulphur respectively. Thus, they should have similar structures. The second statement is more qualitative than the first; but to be precise the bond angle of a tetrahedron i.e. 109.47° ; whereas, the bond angles of the <HOH is 104.5° in water and <HSH in hydrogen sulphide is 92.1°. Elucidation of these small differences in the bond parameters makes more quantitative understanding on the structures. To explain the differences in bond angles in the two compounds, information on the electro-negativity differences between oxygen atom and sulphur atom are essential.

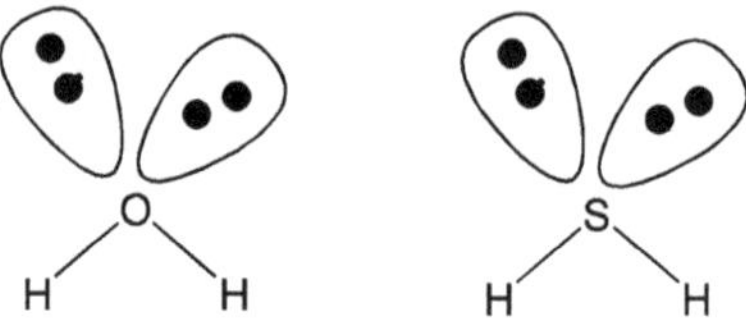

Fig. 1.1 Structure of H₂O and H₂S.

More and more questions can be posed on the structure, property, and reactivity of the water and hydrogen sulphide molecules. The differences in reactivity, physical properties, acidity, and solubility cannot be explained on the basis of one parameter. Each of them requires independent explanations and experimental evidences. Symmetry wise both, water and hydrogen sulphide belongs to same point group C_{2v}. The criterion for this is that the molecules posses a C_2 axis and two vertical planes of symmetry. In these molecules the C_2 axis is through oxygen or sulphur molecules bisecting the lone pairs and the OH or SH bonds respectively. They contain two perpendicular planes, one containing the molecule and other bisecting the H-O-H bonds or H-S-H bonds. The vibrational analysis based on symmetry provides the different vibrational absorptions possible in these compounds; which can be confirmed by recording infrared spectra of these compounds.

The description of discrete structure of water molecules and that of hydrogen sulphide based on covalency is not enough to explain why water is a liquid and hydrogen sulphide is a gas at ordinary condition. For this purpose we have to invoke the hydrogen bonding ability; which is after all a new parameter over the chemical bonds formed by sharing of electrons or by electrostatic forces. The hydrogen bond is a weak interaction, between acidic hydrogen with an electronegative atom in a molecule. Such bond strengths vary and depending on bond strength. They are classified into three categories – weak, medium, and strong. Hydrogen bond interactions can be either inter-molecular or intra-molecular. The context of intra-molecular hydrogen bond is relevant to understand physical properties such as melting point, solubility etc. in particular compounds.

Ionisation Equilibrium

The difference in acidity between water and hydrogen sulphide is another point of attention. Water is a liquid and is neutral; it ionizes as

$$2H_2O \rightleftharpoons H_3O^{\oplus} + OH^{\ominus} \qquad \ldots\ldots(1.1)$$

The equilibrium ($K_w = 1 \times 10^{-14}$ mol L^{-1}) can be studied by shifting it towards right or left.   This can be done by adding acid or base to water. The entire process can be described by ionisation. The ionisation equilibriums are associated with equilibrium constants. Based on the ionisation of acids or bases, hydrogen ion concentration scale is developed. It is called the pH; alternatively other terms such as pK$_w$ and pOH are used to designate the equilibrium concentration and hydroxyl ion concentration. The ionisation processes are established by measuring parameters such as solvent polarity, covalency associated with solute, solvation, long range interactions etc. A prior knowledge on the differences between weak and strong acids helps in choosing the appropriate acid for such studies. The ionisation equilibrium for each acid is different. The criterion of the acid to shift such equilibrium is by studying the ionisation of an acid or by treating the acid with an appropriate base. A strong acid ionizes completely in water. However, a weak acid ionises incompletely and leads to the concept of degree of dissociation. The point to be noted here is that the understanding such equilibrium processes may lead to develop other related concepts. For example let us take a weak acid dissociating as:

$$HA \rightleftharpoons H^{\oplus} + A^{\ominus} \qquad \ldots\ldots(1.2)$$

$$K_a = \frac{[\overset{\oplus}{H}][\overset{\ominus}{A}]}{[HA]}$$

$$.....(1.3)$$

While studying such equilibriums, the concentration of some intermediate species cannot be determined directly. These quantities can be determined indirectly by taking help of a proportionate measurable quantity related to the concentration of the species under consideration. For example, the degree of dissociation can be estimated by the ratio of the conductance offered by the acid at a particular concentration with respect to the conductance at infinite dilution. A strong electrolyte dissociates completely under ordinary condition; thereby the dissociation is unaffected by dilution. But, the dissociation of a weak electrolyte is decided by concentration at which it is studied. As the concentration of electrolyte decreases, it ionizes and completes ionisation at infinite dilution. To confirm this one may put forward that the conductance of a constant volume of solution with different concentration will vary. The conductance is governed by number of ions present in a solution. But this is a qualitative explanation; for quantitative purpose the measurement of specific conductance is necessary. Specific conductance is the conductance, offered by a solution of $1cm^3$ placed between two electrodes; which are separated by 1cm apart. However, to make a general study the concentration of the solution in terms of molarity and normality should appear with conductance. For such purposes the quantity such as molar conductance and equivalent conductance are in use. As mentioned, multiple numbers of equilibriums are associated with di, tri-basic acids depending on their ability to deprotonate in steps. But a common titration experiment performed by using an indicator may not have the scope to distinguish stepwise processes. This is due to the fact that ionisation process itself is fast. The conventional indicator generally changes its color when the complete ionisation takes place. But, it may be possible to understand such equilibriums through conductometric titrations. The conductometric titrations are based on the conductance offered by each ionic species formed in an ionisation process. The other important aspect of such conductometric titration is the end points determined by graphical method; leading to higher accuracy. Let us look at the independent reactions of two different acids namely hydrochloric acid and oxalic acid with sodium hydroxide; the reactions are represented in scheme 1.1.

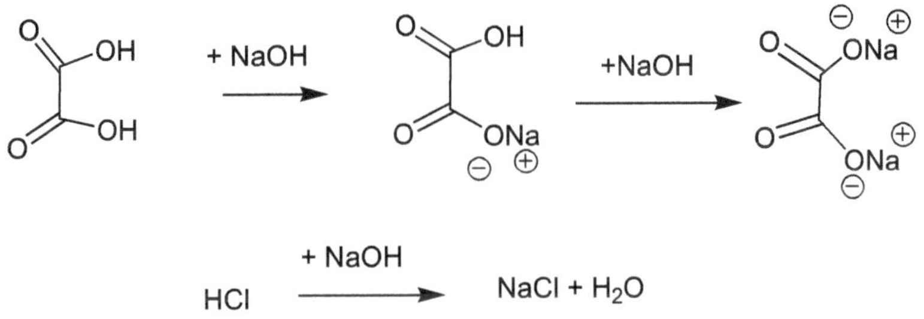

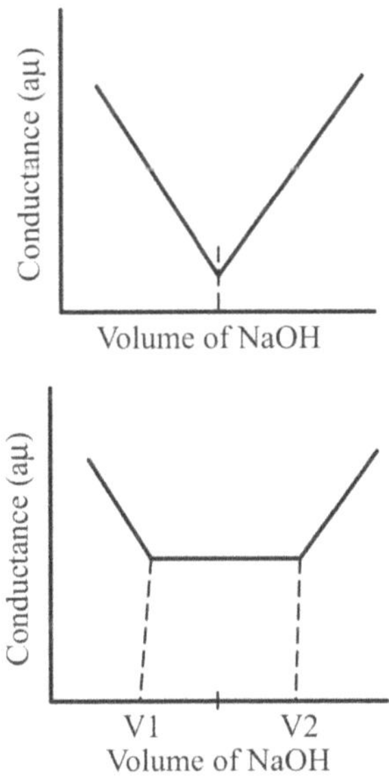

Scheme 1.1 Salt formation in acids.

Titration between these acids with sodium hydroxide can be studied in several ways, a conventional volumetric titration by using phenolphthalein as indicator, conductometric titration, pH titration and potentiometric titration. The conductometric titration is based on the conductance of different ions with different conductivity values. The Kahlaurch law of independent migration states that the total conductance offered by a solution is the summation of conductance of ions at infinite dilution. The plots that would be obtained from the two conductometric titrations are shown in the Fig. 1.2.

Fig. 1.2 Conductometric titration of hydrochloric acid (top) with sodium hydroxide and oxalic acid (bottom).

The plots can be explained in terms of the conductance offered by the ions. In the reaction of hydrochloric acid with sodium hydroxide, the highly conducting H^+ ions are replaced by relatively less conductive sodium ions. In addition to these neutral water molecules are formed. Thus, conductance decreases as the neutralization process proceeds and after reaching the end point excess of sodium hydroxide takes over the conductance. At this stage the OH^- has very high conductance, so stiff rise in conductance is observed. It is clear from the plot shown in Fig. 1.2 (bottom); in the case of oxalic acid the two ionisation process can be seen. After completion of the first ionisation there is slow rise in conductance. When second neutralization takes place there is a sharp increase in conductance.

Alternatively, measuring the pH of known amount of acid solution with increment of added base both these experiments can be performed. The pH is not a directly measurable quantity. It is determined as a function of hydrogen ion concentration; as it is the $-\log$ [H+]. The electromotive force or voltage is measured by a pH meter. A pH meter measures the electro motive force and mathematically transforms to pH. The pH meter has a glass electrode. A glass electrode comprises of silver and silver chloride in 0.1M aqueous hydrochloric acid. The glass electrode is used to measure the pH of the solution. It has also a secondary standard electrode namely calomel electrode. The later is used for calibration of the equipment with a standard buffer solution. A calomel electrode is made of mercury, mercurous chloride in saturated solution of potassium chloride. This serves as standard electrode with reference to a hydrogen electrode. The principal equation involved in pH measurement by a glass electrode is

$$E = \{E_{SHE} + (RT/F) \ln a_H^+\} - E_{calomel} = (-59.15mV)pH - E_{calomel}.$$

These titration experiments can be performed with a potentiometer by measuring the voltage. However, potentiometric titration requires some redox couple to provide redox potential which is absent in the case of an ordinary acid and base. Namely, a cell comprising of two platinum electrodes may be good enough to electrolyse an acid. It cannot provide information on the pH of the solution without an additional redox couple. However, the addition of quinhydrone (an equimolar amount of 1,4-benzenediol and 1,4-benzoquinone) to the solution can provide the requisite redox couple for the same cell to measure pH. The cell will now be sensitive to hydrogen ion concentration to perform a potentiometric titration of an acid and a base. The redox reaction in this process is shown in scheme 1.2. They involve generation of 1,4-

dihydoxybenzene through two stepwise, one electron transfer processes. The later compound formed is a phenolic compound capable of releasing protons easily as shown in the second step of the scheme 1.2.

+e
-e
+ H+
+e
-e
+ 2H+

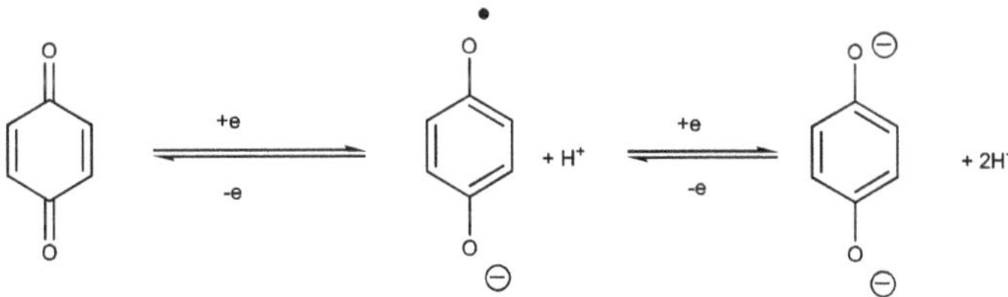

Scheme 1.2 Redox processes in benzoquinone.

The simple acid base chemistry finds application in mineral growth and leads to formation of scenic structures. For example by reaction of water with carbondioxide and calcium carbonate leads to formation of calcium bicarbonate. In limestone cave it drips from top and also grows in crystalline forms calcium carbonate by losing carbondioxide in the form of solid. The droplets hanging from top solidify to form beautiful structures known as stalactite and the one grows from the bottom are called stalagmite (Scheme 1.3).

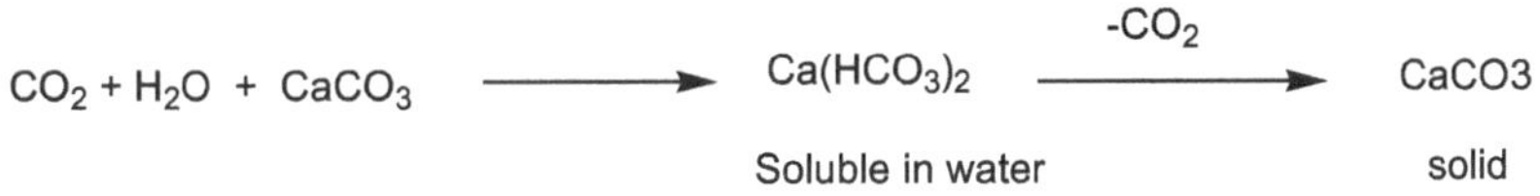

CO2 + H2O + CaCO3
Ca(HCO3)2
Soluble in water
-CO2
CaCO3
solid

Scheme 1.3 Formation of stalagmite and stalactite.

Reaction of water with carbondioxide is industrially important. It produces sodium carbonate as well as sodium hydrogen carbonate. The principle is based on the fact that, sodium carbonate is produced by heating a mixture of sodium hydroxide, carbondioxide and water. Sodium bicarbonate is prepared by mixing sodium carbonate and carbondioxide with water (Scheme 1.4).

CO2 + H2O
NaOH
HCl
CO2 + H2O
mixing
NaHCO3
Heat
Na2CO3
NaOH
HCl
Na2CO3

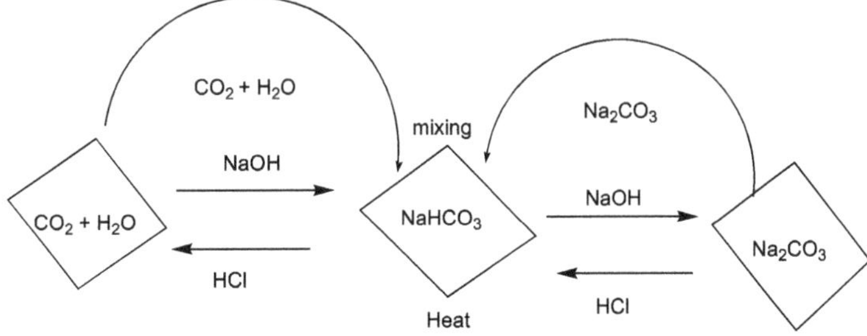

Scheme 1.4 Sodium carbonate synthesis.

Many of the oxy-acids have stable anhydrides, which on hydrolysis gives corresponding acid. Some such reactions are shown in scheme 1.5.

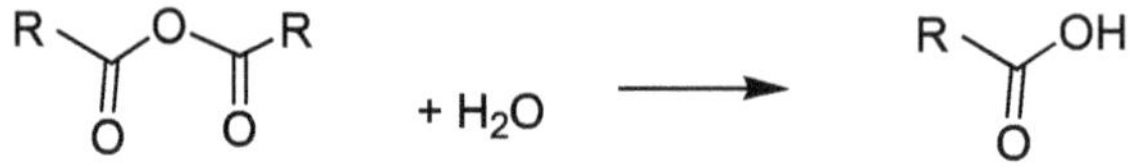

$$CO_2 + H_2O \longrightarrow H_2CO_3$$

$$SO_3 + H_2O \longrightarrow H_2SO_4$$

$$P_2O_5 + H_2O \longrightarrow H_3PO_4$$

$$NO_2 + H_2O \longrightarrow HNO_3$$

Scheme 1.5 Some hydrolysis reactions of acid anhydrides.

Several hydrolytic reactions are so violent that explosion can be caused by such reactions. For example, the reaction of sodium, potassium, and lithium with water can cause explosion.

$$Na + H_2O \longrightarrow NaOH + H_2 \quad \text{explosive reaction} \qquad .....(1.4)$$

Similarly, hydrides react violently with water to liberate hydrogen gas. Advantages of such reactions are thus taken for generation of nascent hydrogen to reduce organic functional groups and also for drying traces of water present in solvents. The reaction of lithium aluminum hydride with water can be fatal. For this purpose to remove traces of unreacted lithium aluminum hydride from reaction mixtures where it is used as a reactant, the reaction mixtures are treated with ethyl acetate, followed by ethanol and then only water is added.

Splitting of water and hydrogen sulphide

Splitting of water by light can be a source for hydrogen and oxygen. When dye D is excited by light, excited state D* is formed. The D* is more oxidizing agent than D.

$$D + h\nu \longrightarrow D^* \qquad .....(1.5)$$

Reacting D* with a quencher Q, D^+ and Q^- can be produced

$$D^* + Q \longrightarrow D^+ + Q^- \qquad .....(1.6)$$

The D^+ and Q^- both can split water in presence of suitable catalyst by reduction or oxidation of water molecules.

$$2Q^- + 2H^+ \xrightarrow{\text{H}_2\text{-catalyst}} 2Q + H_2 \quad \dots(1.7)$$

$$4D^+ + 2H_2O \xrightarrow{\text{O}_2\text{-catalyst}} 4D + 4H^+ + O_2 \quad \dots(1.8)$$

The reactions will depend on the efficiency of catalyst and such reactions will stop if the neutralization of the D^+ and Q^- takes place easily.

$$D^+ + Q^- \longrightarrow D + Q + \text{heat} \quad \dots(1.9)$$

The overall reaction is:

$$2H_2O \xrightarrow[\text{D} + h\nu]{\text{H}_2\,\text{cat} + \text{O}_2\,\text{cat}} 2H_2 + O_2 \quad \dots(1.10)$$

Based on these, cyclic processes as shown in scheme 1.6 can be constructed. In these cycles *tris*-2,2'-bipyridine ruthenium(III) complex and methyl viologen are used as dye (D) and sensitizer (Q) respectively.

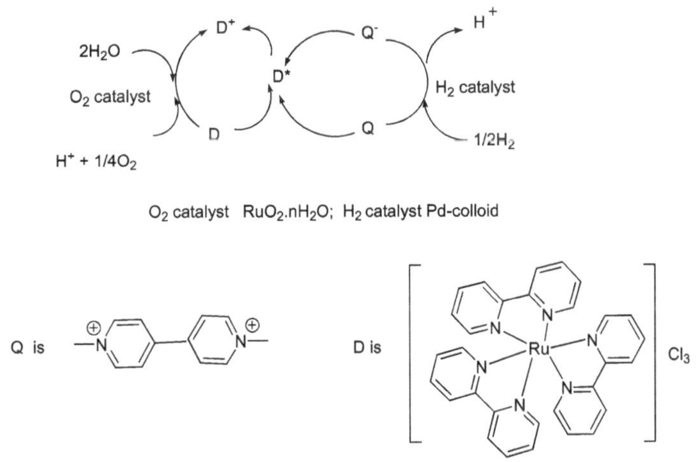

Scheme 1.6 Photocatalytic splitting of water.

Splitting of water to hydrogen and oxygen takes place under photochemical condition. Semiconductors such as titanium dioxide are also catalyst for such reactions. When semiconductor catalysts are used to excite electrons from the valence band to conduction band of a semiconductor, during the process electron holes are produced. These holes facilitate electronic excitation upon irradiation.

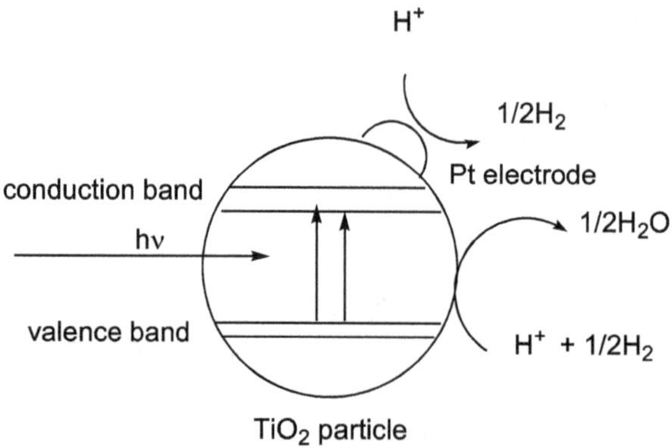

Fig. 1.3 Splitting of water catalysed by semi-conductor.

Radiolysis of water by decay of radioactive isotope ^{60}Co source generates hydroxyl radical. Solvated electrons and hydrogen are formed during such radiolysis. They find application in hydroxylation reactions of organic compounds. The solvated electrons lead to the formation of hydroxyl radical as illustrated in scheme 1.7. Such radicals can generate carbon radicals on sterically crowded alcohol such as t-butylalcohol. However, in the presence of isopropanol the hydroxyl radical leads to isopropyl alcohol radical. Thus, substrates like isopropanol can scavenge the formation of hydroxyl radical in radiolysis.

$$H_2O \xrightarrow{\text{radiolysis}} {}^{\bullet}OH,\ e_{aq}^{-},\ H_2O_2,\ H_2,\ H_3O^{\oplus}$$

$$^{\bullet}OH + (CH_3)_3COH \longrightarrow {}^{\bullet}CH_2(CH_3)_2COH + H_2O$$

Scheme 1.7 Radiolysis of water.

Water in green chemistry

Water is a solvent of choice for green chemistry. Due to government ordinances, concern about the environment, and also to follow the principles of green chemistry, it is important to carry out reactions in aqueous medium. Water soluble reagents are developed to perform reactions in aqueous medium. In green chemistry a set of principles to reduce or eliminate the use or generation of hazardous substances is followed. There are twelve principles, which are as follows:

1. It is better to prevent waste than to treat or clean up waste after it is formed.

2. Synthetic methods should be designed to maximise the incorporation of all materials used in the process into the final product.

3. Wherever practicable, synthetic methodologies should be designed to use and generate substances that possess little or no toxicity to human health and the environment.

4. Chemical products should be designed to preserve efficiency of function while reducing toxicity.

5. The use of auxiliary substances (e.g. solvents, separation agents, etc) should be made unnecessary wherever possible and, innocuous when used.

6. Energy requirements should be recognized for their environmental and economic impacts and should be minimised. Synthetic methods should be conducted at ambient temperature and pressure.

7. A raw material of feedstock should be renewable rather depleting wherever technically and economically practicable.

8. Unnecessary derivatisation (blocking group, protection/ deprotection, and temporary modification of physical/chemical processes) should be avoided whenever possible.

9. Catalytic reagents (as selective as possible) are superior to stoichiometric reagents.

10. Chemical products should be designed so that at the end of their function they do not persist in the environment and break down into innocuous degradation products.

11. Analytical methodologies need to be further developed to allow for real-time, in-process monitoring and control prior to the formation of hazardous substances.

12. Substances and the form of a substance used in a chemical process should be chosen so as to minimise the potential for chemical accidents, including releases, explosions, and fire.

Application of water in day to day Life

The property of water as solvent is universal and versatile. It is consumed by all living beings in the earth. Besides solubility of metal salts, the heat of hydration is an important parameter; it is used in determining several other thermodynamic parameters such as enthalpy, entropy, and free energy. These parameters help in estimating solute-solvent interactions that can be estimated by calorimetry. However, sophisticated equipment like UV-visible spectrometer, fluorescence spectrometer, NMR spectrometer, differential scanning calorimeter are used to understand such properties with higher accuracy.

Hard water and soft water both have their own value. The hard water contains minerals that are good for health. The hard water is bad for cleaning purpose. It leads to formation of calcium and magnesium salts of detergents, which prevents foam formation. The cleaning process involves concept of micelle formation. Soaps or detergents have long chain hydrophobic part and polar head groups such as carboxylate or sulphonate (Fig. 1.4). The polar head groups self-assemble by the interactions with water; are put apart or align the hydrophilic groups. This results in the formation of micelle. The micelles are responsible for removal of the dirt molecules while washing. The micelle formation takes

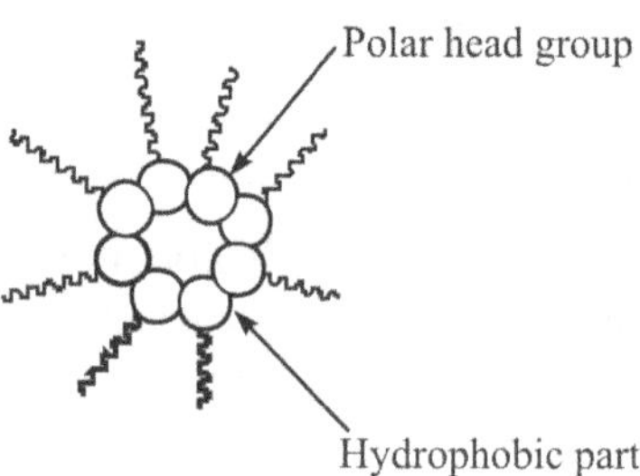

Fig. 1.4 Description of a micelle.

place at a particular concentration and this concentration is known as critical micelle concentration. The formation of micelle is also affected by ionic strength of a solvent. This may arise from the dissolution of impure metal ions or deliberately added metal ions; which controls the ionic strength of water. Use of surfactant in water increases the cleaning ability.

Catalysis

Water as a source of oxygen to form carbon–oxygen bond with relatively inert compounds such as carbon monoxide is industrially important. Water gas shift reaction (equation 1.11) is catalysed by various metal catalysts. This reaction is of special interest as the carbon monoxide a toxic gas can be converted to carbon dioxide at the expenses of readily available water which further results in hydrogen gas. Over all the product can be used as a fuel.

$$H_2O + CO \rightleftharpoons H_2 + CO_2$$

.....(1.11)

Two such catalytic cycles are illustrated in schemes 1.8 and 1.9.

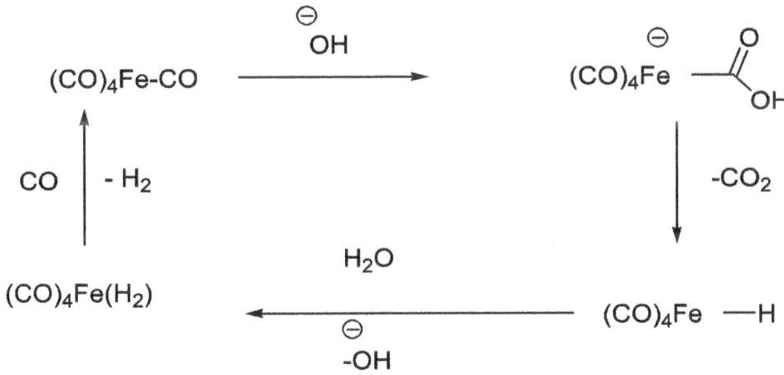

Scheme 1.8 $Fe(CO)_5$ catalysed reaction in basic medium.

Both these reactions involve oxidative addition and reductive elimination reaction as key steps for completing the catalytic cycle. To achieve these conditions in both the cases, low oxidation states metal complexes are used as catalysts. The ligands such as carbon monoxide and phosphines are useful in such reactions. They have back bonding abilities to stabilise low oxidation states of metal ions, thereby help in efficient catalytic process.

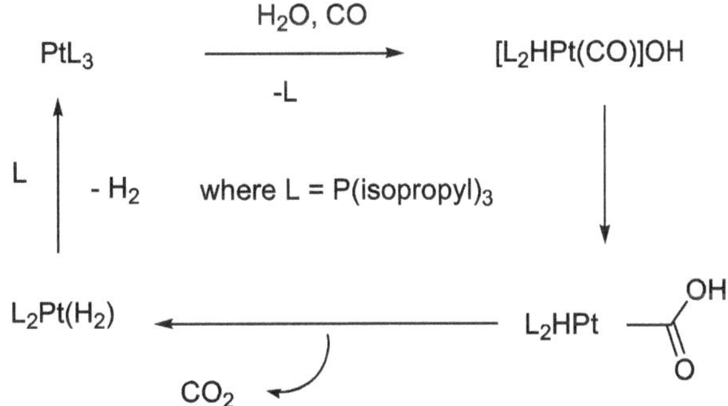

Scheme 1.9 A platinum complex catalysed water shift reaction.

Pollution

Water pollution is studied in great details by environmental chemist. Prevention, remedy for pollution, as well as the hazard caused by water pollution are studied from different point of views. The permissible limit of metal, non-metal content and organic material are determined by various conventional techniques. There is also necessity for determination of microbial and biological pollutant in water. The equipment like ion-selective electrodes, UV-visible spectrometer, fluorescence spectrometer, inductively coupled mass spectrometer, X-ray fluorescence spectrometer, electro-analyser, atomic absorption spectrophotometer, gas chromatograph, high performance liquid chromatography, thermal energy gas chromatograph, auto-analyser for hydrogen cyanide detection are commonly required laboratory equipment for water pollution study. Although these are list of equipment used for analysis of water pollution, but the pollution problems are localized and demographic in nature. For example, the industrial wastes are dependent on the products that are developed. The waste produced by a fabric industry is far different from a waste produced in a petrochemical industry. These examples clearly suggest how analytical skill can be used to collect sets of data but the ability to analyse a chemical problem is essential to arrive at a conclusion. This is irrespective of the domain of fundamental research.

Acid rain refers to rain or any other form of precipitation that is unusually acidic in nature. Acid rain may be a consequence of air pollution. This is toxic for human and any living being. It is caused by release of nitrogen and sulphurous compounds to atmosphere by known or unknown ways. Clean unpolluted rainwater is slightly acidic; has pH of about 5.2; this is due to reaction of water with atmospheric carbondioxide to give carbonic acid. However pollutant like sulphur trioxide generated from oxidation of sulphur dioxide along with nitrogen dioxides gets hydrolysed; they form sulphuric acid and nitric acid respectively. These acid molecules are trapped in the cloud, decreases the pH of the rainwater.

Sometimes rain water also gets contaminated by oxidant such as hydrogen peroxide. Trace amount of hydrogen peroxide in rainwater samples can be determined by fluorescence quenching method. The method is based on the reaction of hydrogen peroxide with 3,3'-diethyloxadicarbocyanine iodide at pH 3.09; the parent compound has emission at 604 nm with excitation at 570 nm. The reaction of 3,3'-diethyloxadicarbocyanine iodide with hydrogen peroxide, leads to compounds which have no fluorescence in acidic solution. Thus, amount of fluorescence quenching can be correlated to the quantity of hydrogen peroxide.

$$\ldots\ldots(1.12)$$

Water of ocean is saline. Salinity of water is the saltiness or dissolved amount of salt in water. The term halinity is used to include the halide content in seawater. The quantity is expressed as parts per thousand. It is convenient to express such a manner, as the quantity tally to grams per liter. However, there is no restriction to express such quantities in gram per million or mg/liter. Salinity decides the type of plants that will grow and makes big impact in ecology of a region. The desalination is to remove excess of salts and minerals from water and to make it ready for human consumption as well as for irrigation. Saline water when irrigated causes increase in salinity of soil. It makes the soil unsuitable for growing crops. The desalination of water by using non-conventional energy source in large scale is a challenge and have prospect of making a desert to an oasis.

Purification of water

Water purification processes are important. Extensive work is needed to make such processes commercial and useful. Thus, this type of chemistry remains contemporary at any time. Adding bleaching powder does purification of water. Bleaching powder has a chemical composition $Ca(OCl)Cl$, it easily liberates chlorine; which is a good oxidizing agent. It is also a disinfectant. The membrane filtration is another common technique for purification of water. Identification of the biocompatible membrane for this purpose is a challenge. Ion-selective separation based on the ion-exchange resins has been widely used in laboratory to prepare de-ionised water. The ion exchange resins can be recovered by treating with an acid or base after it is being used. The ion exchange resins in the form of sulphonate salts of polymers are commercially available in the form of bids. While exchanging cations the resins get converted to sulphonated salt of the cation. The sulphonated salt formed after cation exchange on treatment with a mineral acid gives back the resin. It is reused for further exchange of cations.

The ultraviolet light is used to purify water. Ultra-violet radiation oxidises organics. Ultra-violet radiation upon interaction with water generates hydroxyl free radicals. Free radicals are short-lived and highly

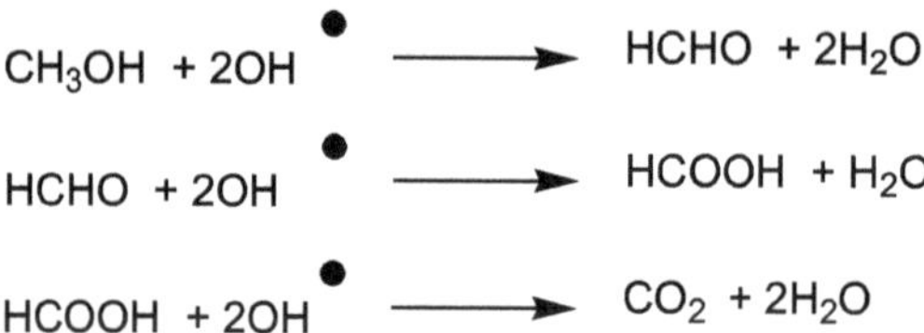

Scheme 1.10 Generation of radical from water.

reactive. They rapidly oxidises many organic and inorganic molecules. The hydroxyl radical (OH•), can be formed via different pathways including the reactions shown in scheme 1.10.

$$CH_3OH + 2OH• \longrightarrow HCHO + 2H_2O$$

$$HCHO + 2OH• \longrightarrow HCOOH + H_2O$$

$$HCOOH + 2OH• \longrightarrow CO_2 + 2H_2O$$

Scheme 1.11 Reactions of hydroxyl radical.

The oxidation by hydroxyl radical can be shown by illustrative example of oxidation of methanol to carbondioxide as shown in scheme 1.11.

Rusting of iron

Rusting of iron is a common phenomenon. Rusting is nothing but corrosion of iron. The rusting process involves reactions of iron with oxygen and water to form iron oxide. Rust consists of $FeO(OH)$, $Fe(OH)_3$, $Fe_2O_3.xH_2O$. In salt water rusting is more efficient. The reactions involved in rusting process starts with the conversion of O_2 to OH^-. During rusting the iron gets oxidized to Fe^{2+} and Fe^{3+}, which lead to iron hydroxides. These hydroxides are dehydrated at different stages to form the rust. Water is indispensable for rusting. The over all reactions involved in the process are shown in scheme 1.12.

$$O_2 + 2H_2O + 4e \longrightarrow 4OH^-$$

$$Fe \longrightarrow Fe^{2+} + 2e$$

$$4Fe^{2+} + O_2 \longrightarrow 4Fe^{3+} + 2O^{2-}$$

$$Fe^{2+} + 2H_2O \rightleftharpoons Fe(OH)_2 + 3H^+$$

$$Fe^{3+} + 3H_2O \longrightarrow Fe(OH)_3 + 3H^+$$

and dehydration equilibrium

$$Fe(OH)_2 \rightleftharpoons FeO + H_2O$$

$$Fe(OH)_3 \rightleftharpoons FeO(OH) + H_2O$$

$$2FeO(OH) \rightleftharpoons Fe_2O_3 + H_2O$$

Scheme 1.12 Rusting process.

Structure of water

A discrete hydrogen bonded assembly of water molecules is called a water cluster. The water clusters are important. They have made diverse impact in chemistry and biology by generating new avenues for structural studies. Water can remain as different aggregates starting from dimer to infinite chain through hydrogen boned interactions. Some of such structures are shown in Fig. 1.5.

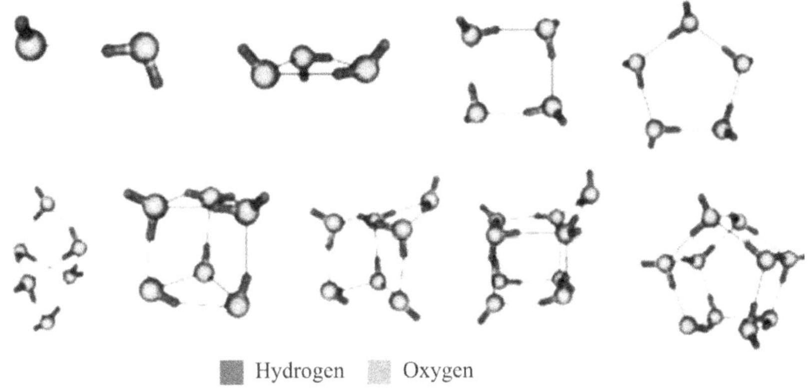

Fig. 1.5 Different hydrogen boned self-assemblies of water.

Experimentally such water clusters are stabilized in different organic and inorganic motifs. For example N-cyclohexyl-2-(quinolin-8-yloxy) acetamide is obtained as pentahydrate. The water molecules around this compound are stablised as tetrameric, hexameric and pentameric water clusters as shown in the Fig. 1.6.

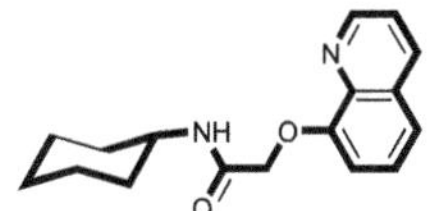

N-cyclohexyl-2-(quinolin-8-yloxy) acetamide

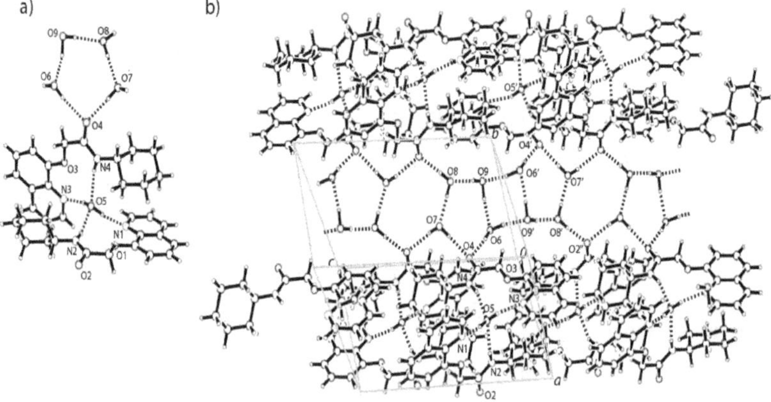

Fig. 1.6 (a) Structure of N-cyclohexyl-2-(quinolin-8-yloxy) acetamide pentahydrate (**1**), and (b) a part of the crystal lattice of **1** showing corresponding hydrogen bonded assembly in the solid-state, featuring tetrameric water clusters which are flanked by hexameric and pentameric water clusters.

Many structural models for structure of water are proposed. The icosahedral structural model with a highly symmetric structure can explain many of the properties of water. An icosahedral structure of water molecule is shown in Fig. 1.7.

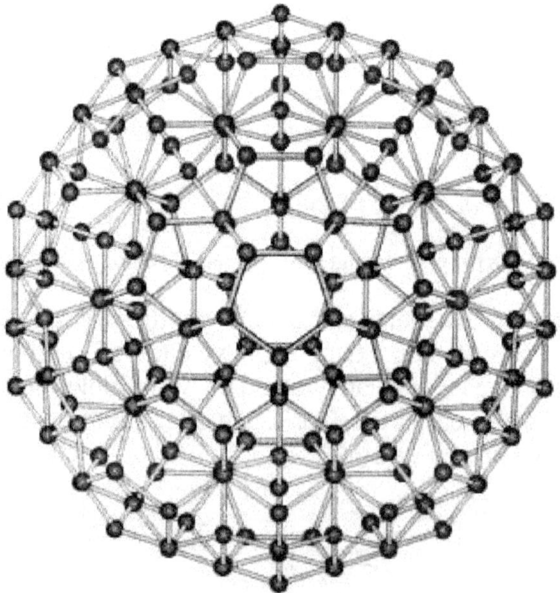

Fig. 1.7 Icosahedral self-assembled structure of water molecules.

Any theory on structure of water must explain anomalous behavior of water molecules. The anomalous properties includes:

(a) Anomalies in physical properties such as melting point, boiling point, critical point, thermal conductivity, faster rate of freezing of hot water than cold water, higher vibration rate of water than cold water, unusually high viscosity and surface tension, prevention of bubble formation by some salts;

(b) Anomalies from water density such as increase in density on heating, shrinking on melting, higher density of surface water than the bulk, reduction in temperature by pressure at maximum density, low coefficient of expansion, increase of thermal expansively with increase in pressure, increase in nearest neighbor on melting, increase in speed of sound above 74°C, Dropping of compressibility after 46.5°C, NMR spin lattice relaxation time at low temperature, maximum refractive index below 0°C, large change of volume while transforming liquid to gas;

(c) Material anomalies, which includes non-ideality of aqueous solution, differences of D_2O and T_2O from water (D = deuterium, T = tritium; both are isotopes of hydrogen), density and viscosity are controlled by solute, high dielectric constant, anomalous ionic mobility, rise of maximum conductivity at 230°C, movement of water molecules far away at high pressure; and

(d) Thermodynamic anomalies such as possession of twice the specific heat of capacity of ice or steam, unusually high heat capacity (C_p and C_v), high heat of vaporization, high heat of sublimation and thermal conductivity reaches maximum at 130°C etc.

Hydration and Polymorphism

The hydration of a material or presence of solvated water in crystal lattice of molecules leads to varieties of compositions. Different solvates of solvated crystals are called pseudo polymorphs or more precisely solvates. In other word, pseudo polymorphism is a phenomenon in which molecules may adopt different solvated forms. For example, a compound may be in the form of monohydrate, dihydrate or trihydrate, in such case this set of molecules are called pseudo-polymorphs. A compound may crystallise with a molecule of solvent such as acetone or ether; in that case both the compounds are pseudo polymorph. The pseudo polymorphism is important in drugs. Some of the hydrated forms of drugs may be active or inactive. Generally a mixture of polymorphic materials makes drugs less active. The drug erythromycin (Fig. 1.8) exists as solvated form of water or acetone. Each of them has different crystal morphology as illustrated in Fig. 1.9.

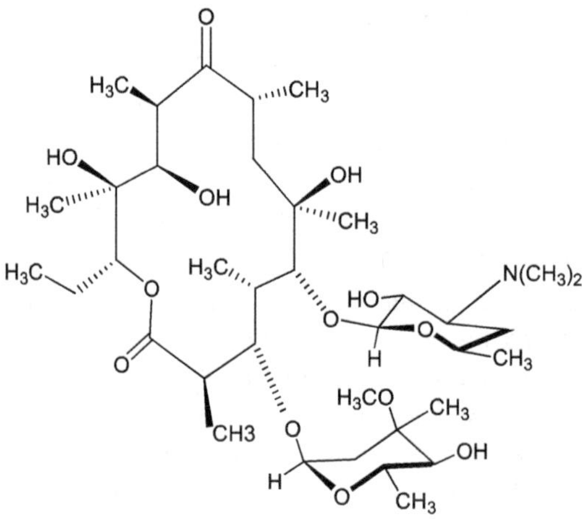

Fig. 1.8 Structure of erythromycin.

Fig. 1.9 Scanning electron micrograph of crystals of (right): erythromycin
dihydrate, (b) erythromycin acetone solvate.

Erythromycin is a macrolide antibiotic. The treatment in human and
veterinary is common with this antibiotic drug. Erythromycin exists as
solvates, anhydrous, and as amorphous solid. But in markets
erythromycin is usually available as the dihydrate. Erythromycin on
crystalisation from acetone crystallises as acetone solvate. Erythromycin
acetone solvate transform to dihydrate form in pure water.

The hydration of a material causes loss or enhancement of fluorescence properties. For example, the fluorescence property of solvates of 2-(9-anthryl) phenanthroimidazole (Fig. 1.10) with water and ammonia are different. It is attributed to the fact that in each case the aromatic stacking pattern are different leading to difference in optical properties.

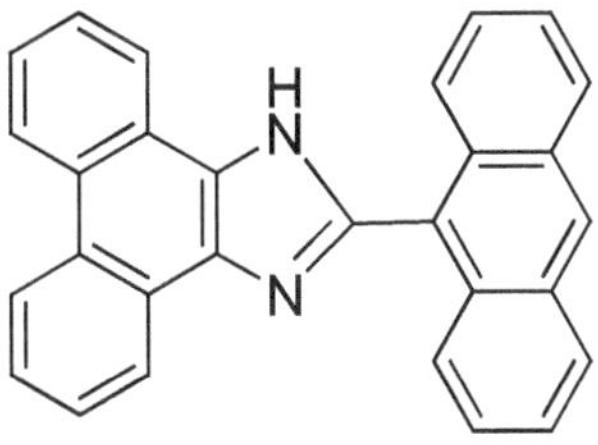

Fig. 1.10 Structure of 2-(9-anthryl) phenanthroimidazole.

The hygroscopic compounds are deliquescent and some of the less stable compounds decompose slowly on contact with moisture. Hydration is a common observation in inorganic salts. For example a bottle of anhydrous ferric chloride left exposed, absorbs rapidly moisture, and can finally look like a concentrated solution of ferric chloride in water. Upon drying it leads to decomposed product, however ferric chloride can be recovered from such mixture upon concentrating/ evaporating the decomposed material after treating with concentrated hydrochloric acid. The evaporation leads to ferric chloride in hydrated form. Preparation of anhydrous ferric chloride requires reaction of the hydrated form with dehydrating agent such as thionyl chloride. Thus, dehydrating agent development becomes very important. The dehydrating agents based on molecular sieves are very attractive. Other dehydrating agents used are phosphorous pentoxide, anhydrous sodium sulphate, anhydrous calcium chloride, anhydrous magnesium sulphate, concentrated sulphuric acid. Some chemical reactions are violent with water, such as reaction of water with alkali metal with water can be used for useful purposes. Traces of water from solvents like, benzene, toluene, tetrahydrofuran, diethylether etc. are removed by reacting with sodium metal. Silica is used as desiccant with color indicators that shows different colour such as green in anhydrous form whereas pink in hydrated form. The indicators used in these are salts of transition metals such as copper, nickel or cobalt.

A dehydration process can be highly exothermic; for example the reaction of water with concentrated sulphuric acid. The heat of hydration can be easily determined from this experiment by using a simple setup comprising of thermally insulated beaker with a thermometer and a stirrer. The thermal insulation can be provided by thermo-cool which is generally used for packing materials. The dehydration of sugar or any carbohydrate by sulphuric acid leads to formation of carbon. The cellulose can be dehydrated by concentrated sulphuric acid. For example, a filter paper turns black on treatment with concentrated sulphuric acid. Dehydrating agents are very useful in protecting damage of electronic items from humidity. In camera lenses one of the common problems is formation of fungus. This makes the lenses opaque when the lenses are kept in humid environment. Protection of such is achieved by keeping small packets of desiccants in the cover where it is placed. Further to this many food items requires prevention of humidity and desiccants provides a means of storage. Drying, dehydration as well as hydration of polymeric materials is very important in fabric industry, as such processes control quality and biological compatibility.

The term azeotrope refers to formation of a uniform mixture with another solvent to have a common boiling point. Only some solvents can form azeotrope with water, common examples are benzene-water, toluene-water azeotropic mixture. This property provides an indirect method to dehydrate a solvent contaminated with water. It also helps in shifting equilibrium of a reaction. For example, acid catalysed ester formation reaction is reversible and can be represented by the equilibrium, shown in equation 1.13.

$$RCOOH + R_1OH \xrightleftharpoons{H^+} RCOOR_1 + H_2O \qquad .....(1.13)$$

By carrying out this reaction in benzene the water formed during the reaction can be removed. Benzene forms azeotrope with water; water formed can be easily removed from the reaction mixture by specially designed apparatus (Dean-Stark's apparatus). Removal of water from the reaction mixture pushes the equilibrium towards right side and esters are prepared in good yield.

Water molecules can be trapped in channel like structures of zeolites. Zeolites are aluminosilicates with a general composition $M^{n+}_{x/n}[Al_xSi_yO_{2x+2y}]^{x-}.zH_2O$ (where x, y, z, n are integers, M = alkali or alkaline earth metal ion). They posses rigid but uniform porous structures. Channel like structures can be formed in commonly available chemicals via weak interactions. A water channel formed by self-assembly of *bis*-phenols is shown in Fig. 1.11.

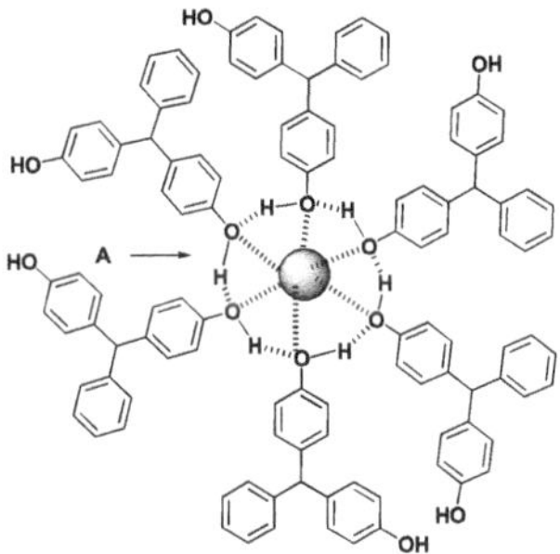

Fig. 1.11 A water channel formed by *bis*-phenol.

The *bis*-phenol has a T-shaped geometry and one each of the hydroxyl groups of six *bis*-phenol molecules can hydrogen bonded via a molecule of water to form a cyclic hexameric network "A". This results in the formation of channel like structure. The channel has diameter of 5.6 Å, where water molecules are trapped. In the channel water molecules are tightly held. From thermogravimetry it is found that the molecules in the channel can be removed at 160 °C, which is much higher than the boiling point of ordinary water.

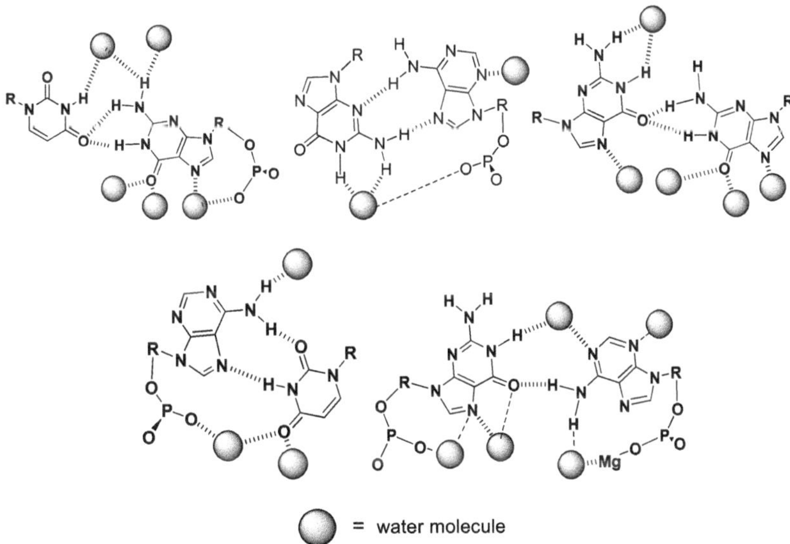

Fig. 1.12 Different base pairs stabilized in different orientations by water
molecule observed as segments in biological system.

Water molecule may help in stabilization of different orientations and arrangements of base pairs of ribonucleic acid. This is illustrated below with the non-canonical base pairs found in the structure of bacterial loop E of 5 S rRNA in Fig. 1.12. Hydrogen bonds are shown with dotted lines.

Surface phenomena

Water bubbles generate water-membrane. These membranes have thickness of the order of nm. Foreign molecules such as dye can be put in these membranes and their movement and their properties in such confined medium are studied. Such processes have relevance to ion transport in biological membrane.

A rainbow is a commonly observed optical phenomenon. It is the observation of spectrum of light in the sky when sun shines on droplet of moisture in the atmosphere. The light spectrum takes the form of a multicolored arc. The light range starting from violet to red spreads up in this arc. As the light enters a droplet of water, it is first refracted and then reflected from the back of the water droplet at an angle ranging from 40°– 42°. The refraction of light is governed by wavelength, for example blue light is refracted at a greater angle than red light. The sunlight has the combination of all the seven colors, i.e. it has mixtures of rays with different wavelengths. Thus, on refraction the colors are separated to show a bunch of colors (Fig. 1.13).

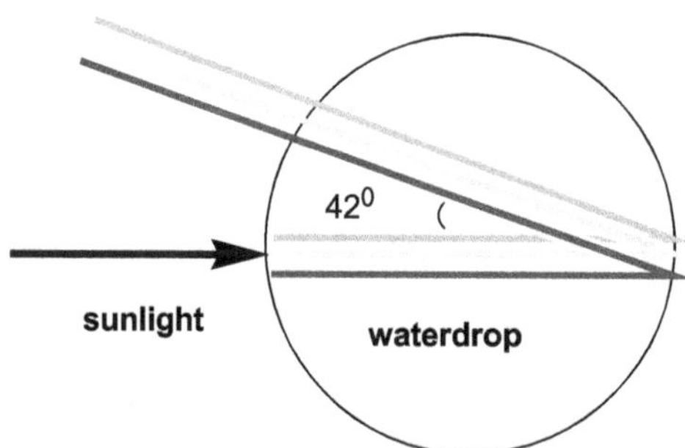

Fig. 1.13 Refraction of light by water bubble.

Capillary action to self-assemble macroscopic object is assisted by water. The polydimethylsiloxane is a polymer which does get wet by water. It can be wetted by fluorinated hydrocarbons. Cut pieces of various dimensions of polydimethylsiloxane in fluorinated solvent on oxidation becomes wettable by water. These oxidized objects float at the interface of perfluorodecalin ($C_{10}F_{18}$) and water. When such assemblies are allowed to move freely, non-oxidised surfaces approached each other within a distance of 5 mm. They move to make contact with each other.

As the time progresses ordered structures are formed as illustrated in Fig. 1.14. Such solvent assisted movement is due to capillary action of water. This is an example ordered self-assembling of macroscopic objects. Similar processes are applicable to molecular level, and have great relevance in biology. Study of such microscopic phenomena requires high-resolution microscope such as transmission electron microscope. Their building up process and time dependent study helps in understanding bottom-up constructions of nano-dimensional materials.

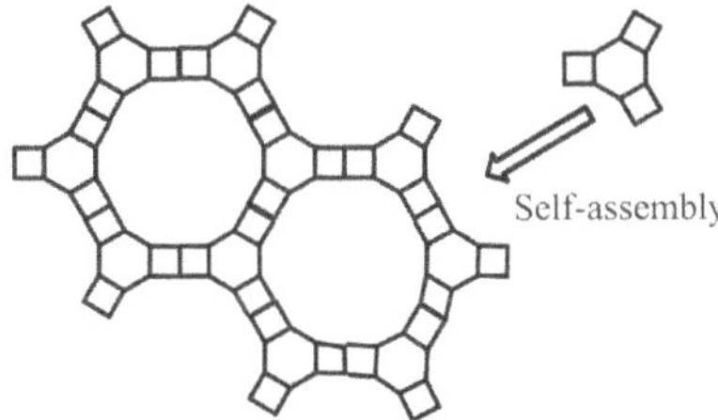

Fig. 1.14 Self-assembling of a partially oxidized polysiloxane at the interface of water and fluorohydrocarbon

Certain properties such as vapor pressure, melting point, freezing point, osmotic pressure, etc are dependent on number of species, which is known as colligative property. As a consequence of such properties the vapor pressure of water changes. This causes depression of freezing point, elevation of melting point. When sodium chloride is dissolved in water, melting point of water is lowered. The freezing point of a 1M sodium chloride solution is roughly $-3.4°C$ at atmospheric pressure.

The freezing point depression occurs because the concentration of water molecules in a solution is less than the concentration in pure water. The process can be better understood from the phase diagram shown in Fig. 1.15. In the phase diagram of water-sodium chloride system, there are clear boundaries for different phases of ice and salt solution. As ice begins to freeze from the solution of salt in water, the fraction of water in the solution becomes lower and the freezing point drops further. This freezing process of ice does not continue indefinitely, because the solution becomes saturated with salt after a particular concentration. The lowest temperature possible for liquid salt solution is $-21°C$. At this temperature, the salt begins to crystallize out of solution as dihydrate of sodium chloride ($NaCl.2H_2O$), along with the ice. This continues until the solution completely freezes. The frozen solution at this stage is a mixture of crystals of sodium chloride dihydrate and ice crystals. It is not a homogeneous mixture of salt and water. This heterogeneous mixture is called a eutectic mixture.

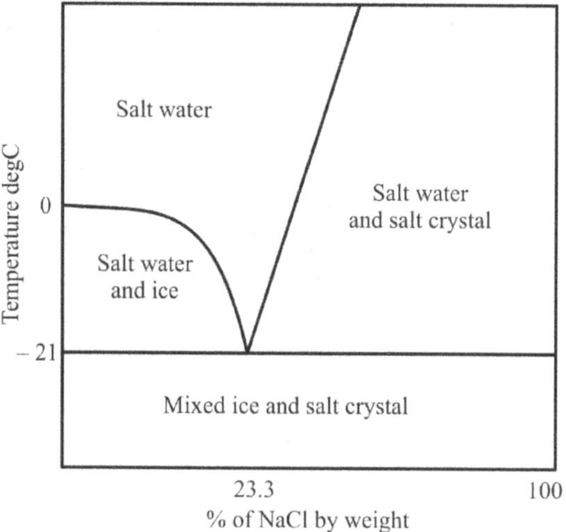

Fig. 1.15 Phase diagram of water-sodium chloride system.

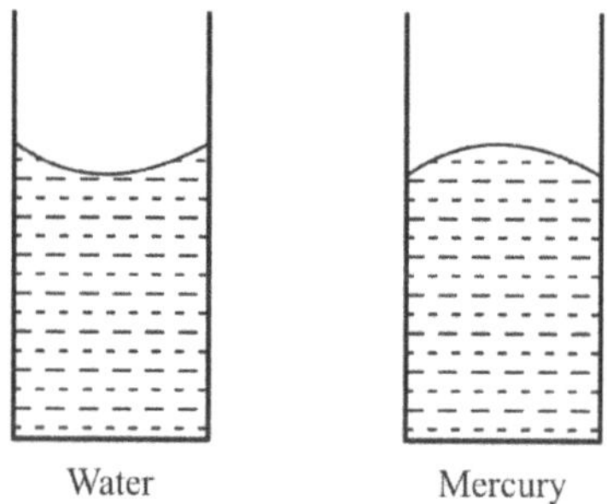

Fig. 1.16 Figure showing shape of meniscus of two liquids.

Water kept in a cylindrical vessel has a concave shape of the meniscus whereas in the case of mercury it appears as convex shape (Fig. 1.16). This is due to the difference in surface tension. Thus, it is the lower meniscus, which is taken as the height of water in a measuring cylinder or a burette used for conventional quantitative estimation.

Some interesting reactions of water

Water is a reactant in hydrolysis reactions and hydrolysis is a very important process in chemistry and biology. For example, an ester can be either hydrolysed by water under ordinary or catalytic condition. Mercury ion assisted addition of water to unsaturated organic compounds find important place in organic chemistry.

$$CH_3CO_2CH_3 \ + H_2O \longrightarrow CH_3COOH \ + \ CH_3OH$$

$$.....(1.14)$$

Water reacts with orgnanometallic compounds (Scheme 1.13). Most of the organometallic reactions are carried out and products are stored in the absence of water, partly due to such reasons.

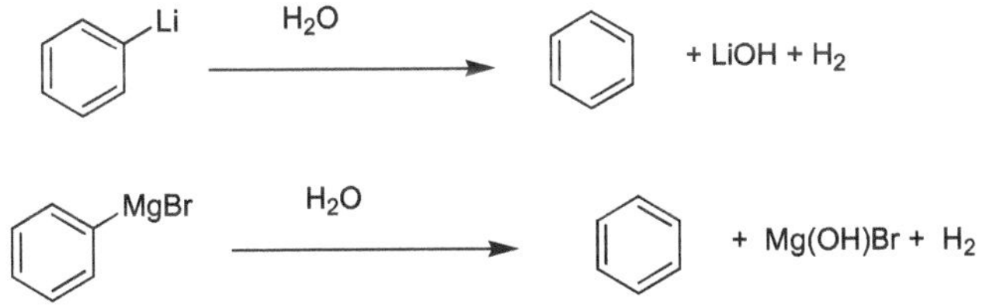

Scheme 1.13 Reaction of metal-carbon bond with water.

Water enriched with isotopes, like deuterium and tritium are useful for labeling experiments. For example, proporgyl alcohol selectively can be converted to a deuterated allylic alcohol by reaction of deuterated water on a lithiated intermediate. The reaction is shown in the scheme 1.14.

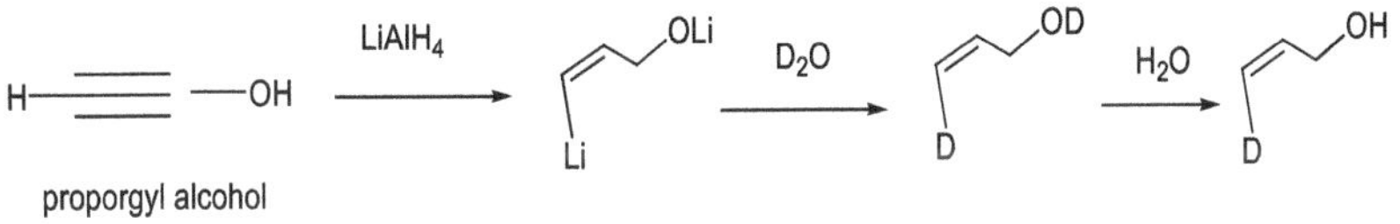

Scheme 1.14 Deuterium incorporation to organic substrate.

Water in biology

The role of water in biology is tremendous. It provides a medium for whole lot of biological processes to the control of biological activities. Hydrolytic cleavage of adenine tri-phosphate leads to adenine di-phosphate generates energy in biological systems. This reaction is one of the important reactions to generate energy within living beings (equation 1.15).

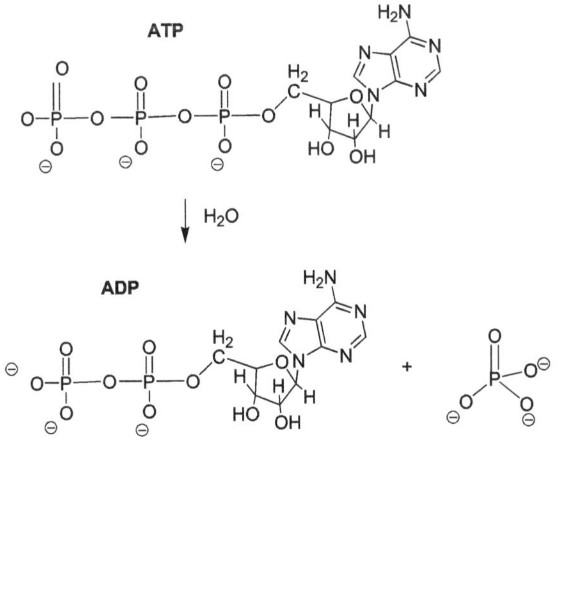

$$.....(1.15)$$

The aqua bridged transition metal complexes are found in nature in active sites of enzymes. For example, active site of methionine amino peptidase of *P. furiosus* has an aqua bridged dinuclear cobalt center, as illustrated in the Fig. 1.17.

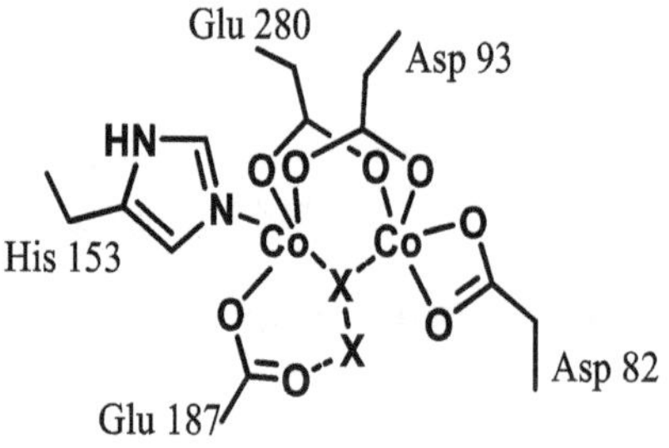

Fig. 1.17 Structure of the binuclear cobalt(II) active site in methionine amino peptidase of *P. furiosus* (X = water and Glu, Asp, His are amino acid residues).

Most of the enzyme activity is controlled by water. Denaturation of any biological species can be caused by removal of water. Quaternary structures of enzymes are due to the hydrophilic and hydrophobic effect of biomolecules in water. Phospholipids form lipid bilayers under the influence of water. These bilayers in turn form vesicles in living body, which plays important role as channels for ion conduction.

Coordination chemistry of water

Water is an important ligand in coordination chemistry. The monodentate and bridging modes are the principal modes of coordination of water to a metal ion. Mechanism of ligand substitution reactions are generally studied with metal aqua complexes. The liability and inertness of inorganic complexes are determined from such studies. For example the rate of ligand exchange reaction of $[Cr(NH_3)_5(H_2O)]^{3+}$ and $[Cr(NH_3)_5(H_2O)]^{2+}$ are different. Chromium(III) having d^3-electronic configuration in a strong octahedral field has all the electrons in t_{2g} levels are equally distributed. Whereas the chromium(II) has $t_{2g}^3 e_g^1$ configuration is labile, such reactions help them to ascertain the electron transfer reaction mechanism. For example, the line broadening in ^{17}O nuclear magnetic resonance spectroscopy is used to determine the rate constant of water exchange of paramagnetic ions.

$$[M(H_2O)_n]^{m+} + xH_2^{17}O \longrightarrow [M(H_2^{17}O)_n]^{m+} + x\,H_2O$$

where m,n,x are integers

.....(1.16)

In these experiments advantage of the differences in the chemical shift in nuclear magnetic resonance spectroscopy of bulk and coordinated water are taken. The line width at the half height of a NMR signal is used to calculate the rate of exchange. Study on ligand exchange reactions between $[Fe(H_2O)_6]^{2+}$ with $[Fe^*(H_2O)_6]^{3+}$ (where Fe* is isotopically labeled iron) is useful to explain the inner-sphere electron transfer mechanism.

The aqua complexes are used not only to explain reactivity but also many of the fundamental properties; color and magnetic properties of such complexes helps in understanding of their origin. For example, hexaaqua manganese(II) complexes are formed on dissolution of different manganese salts in water. The $[Mn(H_2O)_6]^{2+}$ has ground state $^6A_{1g}$ and should not show any d-d transition. Thus, it should show a colorless solution. However, these species are light pink. It occurs due to coupling between vibrational levels with electronic levels. Such mixing of vibrational level with electronic level is known as vibronic coupling.

Some properties of hydrogen sulphide

The hydrogen sulphide is produced in laboratory by reaction of Al_2S_3 with water or FeS with hydrochloric acid. Many bacteria produce hydrogen sulphide by reduction of sulphate containing compounds. Some anaerobic bacteria produce hydrogen sulphide when they digest sulphur containing amino acids. Some bacteria in human colon produce hydrogen sulphide and odor of flatulence can be due to this. Some bacteria in mouth also can produce hydrogen sulphide leading to bad smell. Hydrogen sulphide is toxic and its high dose can cause serious health problem. However, low limit of hydrogen sulphide can be tolerated by human being. In human body there are enzymes capable of detoxifying hydrogen sulphide by oxidation to harmless sulphate. A highly concentrated hydrogen sulphide environment leads to discolorisation of silver coin.

The hydrogen sulphide dissociates as

$$H_2S \longrightarrow HS^- + H^+ \qquad\qquad .....(1.17)$$

The K_a value is 6.89 X 10^{-7} mol/lit with a pK_a 6.89. However, second dissociation in the case of hydrogen sulphide does not occur in solution; it has a pKa$_2$ 14.18. The S^{2-} dianions are known to form metal sulphides which are generally solids. Several of such metal sulphides have interesting material properties; for example, cadmium sulphide is a semiconductor. The hydrogen sulphide at low temperature solidifies as hydrates which has a composition $H_2S.5.75H_2O$.

Hydrogen sulphide finds important place in cation analysis. It is useful for precipitating some metal salts at different pH conditions. In all the cases of cation analysis water is used as medium. We donot consider the role of water in these reactions generally act as solvent. The focus of attention in quantitative analysis is on the precipitation and separation of the cations. The precipitation reactions of metal ions are based on the following equilibriums:

$$H_2S + H_2O \rightleftharpoons HS^- + H_3O^{\oplus} \qquad K_1 = [H^+]\{HS^-\}/[H_2S]$$

$$HS^- + H_2O \rightleftharpoons S^{2-} + H_3O^{\oplus} \qquad K_2 = [H^+]\{S^{2-}\}/[HS^-]$$

$$H_2O \rightleftharpoons H^+ + OH^-$$

Scheme 1.15 Deprotonation equilibriums of hydrogen sulphide.

To maintain overall neutrality total cation concentrations must be equal to anion concentration. So,

$$[H^+] = [HS^-] + 2[S^{2-}] + [OH^-] \qquad \qquad .....(1.18)$$

While dealing with acidic solution $[H^+] > 10^{-7} > [OH^-]$; one can write

$$[H^+] = [HS^-] + 2[S^{2-}] \qquad \qquad .....(1.19)$$

If one considers a x-molar solution of hydrogen sulphide

$$[H_2S] + [HS^-] + [S^{2-}] = x \qquad \qquad .....(1.20)$$

Generally K_2 is small, therefore at very dilute concentration of hydrogen sulphide solution, the concentration of the $[S^{2-}]$ is also very less and ignoring this term in equation 1.17 one may write $[H^+] = [HS^-]$, since K_1 is also a small quantity, $[H^+] << [H_2S]$ hence equation 18 may be reduced to $[H_2S] = x$. From these sets of equations, the total sulphide, hydrogen sulphide and hydrogen ion concentration are determined. Based on the above equilibriums, the competitive precipitations of different metal sulphides at different hydrogen ion concentrations are evaluated.

These exemplify the fact that the reactivity of two similar compounds (in this case water and hydrogen sulphide) vary widely. The appropriate competitive reactivity is to be ascertained before performing reactions which have similar features. Hydrogen sulphide is a biproduct of oil or natural gas. Its application to generate energy is important. It also requires much lower energy in comparison to water for its dissociation to hydrogen and sulphide. Irradiation of light to semiconductor produces electron-hole and the reduction of hydrogen sulphide takes place similar to water but in the case of hydrogen sulphide solid sulphur is formed rather than gaseous oxygen that is formed from water. Thus, waste management issue makes it less attractive.

Chemistry with Ammonia and Borane

General discussion

From the foregoing discussions, it is clear that the different types of experiments are needed to understand different phenomenon arising either at molecular level or at bulk. A similar discussion with another pair of molecules, borane and ammonia would exemplify the need of understanding chemistry through experiments. The two molecules borane and ammonia are tetra-atomic molecules but they do not have similar structure. Borane is planar molecule and ammonia has a pyramidal structure. Are these good enough to explain the lewis acidity or basicity of borane and ammonia independently? The borane forms dimer easily to give B_2H_6, which has three center two electron bond. The three center bond is formed by two overlap of two sp^2 hybrid boron atoms with a 1s orbital of hydrogen as illustrated in Fig. 1.18. The diborane may be considered to be formed through the overlapping of H-orbitals one each from two B-H bond of two independent boranes with the vacant orbital of boranes. These orbitals have similar energy and also proper orientations to have appreciable spatial overlap. In diborane, each boron atom is associated with two different types of bonds. One set is 3c-2e B-H-B bond and the others are two conventional 2c-2e B-H bonds. In case of ammonia there are four sp^3 hybrid nitrogen orbitals one accommodates the lone pair and the other three the three N-H bonds.

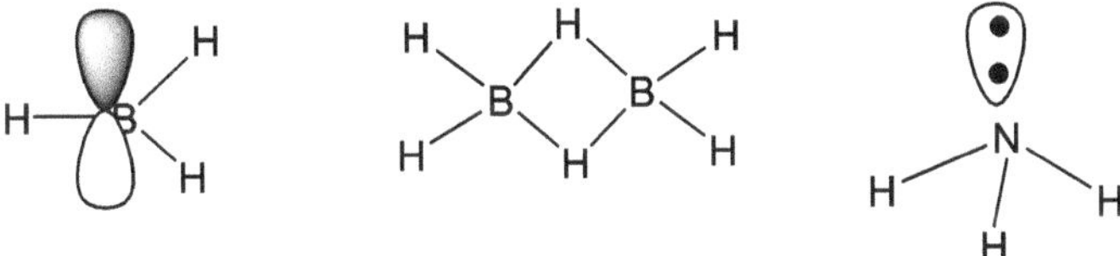

Fig. 1.18 Structure of borane and ammonia

The electronegativity difference between boron and nitrogen is not good enough to explain the differences in the reactivity of ammonia and borane. It is obvious that these two pairs are not isoelectronic. Thus, the electron counting becomes more essential in this case to arrive at the structural differences between these two compounds. Differences in electronic configurations also explain the reactivity. The pyrolysis is B_2H_6 is extremely complex; it leads to various boranes such as B_3H_9, B_4H_{10}, B_5H_{11} etc (Scheme 1.16).

$$B_2H_6 \rightleftharpoons 2\{BH_3\}$$

$$\{BH_3\} + B_2H_6 \rightleftharpoons B_3H_9 \xrightarrow{\text{rate determining}} \{B_3H_7\} + H_2$$

$$\{BH_3\} + \{B_3H_7\} \rightleftharpoons B_4H_{10}$$

$$B_2H_6 + \{B_3H_7\} \longrightarrow \{BH_3\} + B_4H_{10} \rightleftharpoons B_5H_{11} + H_2$$

Scheme 1.16 The equilibriums between different boranes.

The structure of the diborane is established with the help of chemical reactivities or with spectroscopic evidences such as mass spectroscopy, nuclear magnetic resonance spectroscopy etc. The B_2H_6 can react with six equivalents of methanol to replace the entire B-H bonds to form new B-OCH_3 bonds. However, in diborane the reactivity of bridging hydrogens are different from the terminal B-H bonds. The core formed by bridging B-N-B bonds reacts with two equivalents of CH_3OH in a slow rate than other the B-H bonds to form two molecules of $B(OCH_3)_3$. However, to distinguish these steps careful experiments are to be performed. The B_2H_6 can be symmetrically or asymmetrically cleaved. These reactions are decided by the steric factors. For example the unsymmetrical products are obtained from NH_3, $MeNH_2$, whereas Me_3N gives symmetric product.

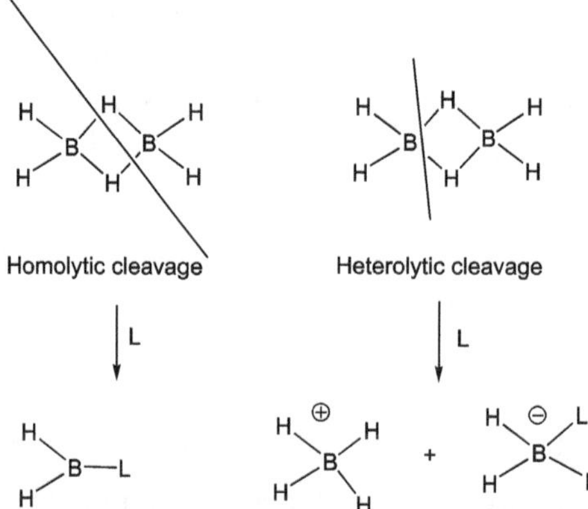

Fig. 1.19 Cleavages of a diborane.

Borane forms a large numbers of clusters. There are three classes of boranes called closo, nido and arachno. These compounds are structurally important. They provide a systematic basis for theoretical chemistry. The theoretical prediction also helps in synthesis of new types of predicted compounds. Based on isolobal analogy B-H fragment with different organometallic or carbon fragments, many organonmetallic complexes and carboranes are prepared.

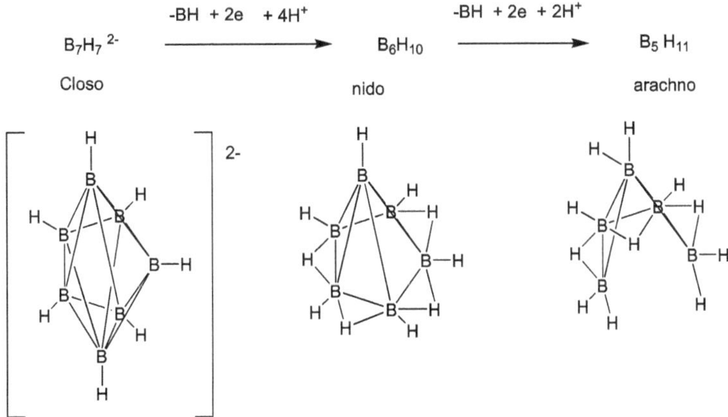

Fig. 1.20 Different types of structures in boranes.

Some reactions of ammonia

The synthesis of ammonia from nitrogen and hydrogen is an illustrative example to apply the concept of catalyst and promoter to a gaseous equilibrium. The equilibrium of this reaction can be shifted to right or left by controlling pressure and temperature.

$$N_2 + 3H_2 \rightleftharpoons 2NH_3 \qquad \dots(1.21)$$

Ammonia is an industrially important compound. The acid base chemistry of ammonia covers a central point of the subject. Ammonia dissolves in water and forms ammonium hydroxide is widely used as base. The K_b the base dissociation constant for ammonia is 1.8×10^{-5} mole/lit.

$$NH_3 + H_2O \rightleftharpoons NH_4^+ + OH^- \qquad \dots(1.22)$$

$$K_b = \frac{[NH_4^+]\,[OH^-]}{[NH_3]} \qquad \dots(1.23)$$

The ammonium salts are easily formed by reactions with different acids and each of them has their own industrial importance. Ammonium salts are used for preparation of buffer solution. The substituted ammonia derivatives, such as primary amine, secondary amine, tertiary amine, quaternary ammonium salts, and amides form a major portion of organic chemistry.

The most important use of industrial use of ammonia is for the production of nitric oxide. Nitric oxide is formed when ammonia is heated along with oxygen over platinum gauze at 850°C (equation 1.24).

$$4NH_3 + 5O_2 \rightarrow 4NO + 6H_2O$$

$$.....(1.24)$$

Ammonia shows some interesting solubility feature. In the case of water the reaction of barium chloride with silver nitrate leads to precipitation of silver chloride. The silver chloride is sparingly soluble in water and barium chloride is soluble in water. However, the barium chloride is insoluble in ammonia but silver chloride is soluble in ammonia. Thus the reaction between barium nitrate and silver nitrate is carried out to selectively precipitate barium chloride.

$$AgNO_3 + BaCl_2 \xrightarrow{H_2O} AgCl\ (ppt) + Ba(NO_3)_2$$

$$.....(1.25)$$

$$Ba(NO_3)_2 + AgCl_{(soluble)} \xrightarrow{NH_3} AgNO_3 + BaCl_{2\ (insoluble)}$$

$$.....(1.26)$$

Ammonia as a solvent leads to interesting solvolytic reactions, for example hexammine cobalt(III) cation reacts with ammonia to form a ammonium cation through deprotonation of a amine ligand:

$$[Co(NH_3)_6]^{3+} + NH_3 \longrightarrow Co(NH_3)_5(NH_2)]^{2+} + NH_4^+$$

$$.....(1.27)$$

Generally the alkali metals are dissolved in water with violent reactions, however the reaction of ammonia with alkali metals are not violent. The dilute solutions of alkali metals in ammonia are blue in color having strong absorption in the near infra red region close to 1500 nm. This absorption position is independent of the alkali metal under consideration. This suggests that it occurs due to a common species. In fact in these cases the blue coloration occurs due to formation of free electrons. The free electrons get solvated by liquid ammonia to form e^-_{am} species (subscript am means solvated ammonia). The metal ammonia solution has high conductivity of electricity. The system has interesting electrical properties as the concentration of metal increases the molar conductance first decreases and on further increase of amount of metal the conductance sharply increases. This is attributed to the formation of metallic ion pair $(M^+_{am})(e^-_{am})$ ionpair. The magnetic susceptibility measurement shows that the blue solution is paramagnetic. The change of conductance as well as magnetic susceptibility changes suggest that there are four types of species contributing to such properties. The formation of ammoniacal electron, pairing of ammoniacal electron, formation of doubly charged paired electrons $[(2M^+_{am} + (e{-}_{am})\ (M^+_{am})_2(e^-_{am})_2]$ and metallic ion pair electron completes these four species.

Ammonia acts as a reducing agent for reduction of calomel to mercury. An amido complex is formed in the reaction which is highly insoluble.

$$Hg_2Cl_2\ +\ 2NH_3 \longrightarrow Hg\ + HgCl(NH_2) + NH_4^+\ + Cl^- \qquad .....(1.28)$$

Derivatives of ammonia are used as base. Ammonia is also used as ligand to prepare different co-ordination complexes. The complex of ammonia with silver, $[Ag(NH_3)_2]^+$ is the active component of Tollen's reagent. This reagent is used to detect carbonyl compounds. The silver mirror is formed on reaction with aldehydes with Tollen's reagent due to reduction of Ag^+ ion to metallic Ag. The tri-ammine tri-chloro chromium(III), $[Cr(NH_3)_3Cl_3]$ exists as two isomers namely facial (fac) and meridonial (mer) isomers (Fig 1.21).

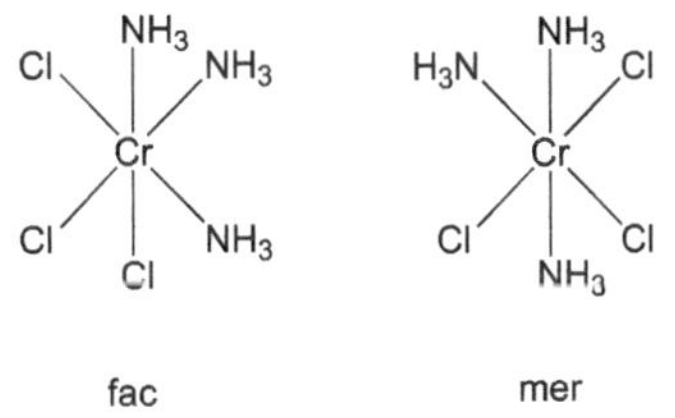

Fig. 1.21 Facial and meridonial isomers.

The lone pair of electrons present on nitrogen atom of ammonia provides umbrella shaped structure to ammonia with respect to the three N-H bonds. The ammonia molecule undergoes inversion, which is commonly known as an umbrella inversion. The energy barrier between the two states is 24.7 KJ/mole. The rate of inversion is very high and can not be detected by conventional methods. Cooling the system to low temperature by using liquid nitrogen or liquid helium such effect is studied.

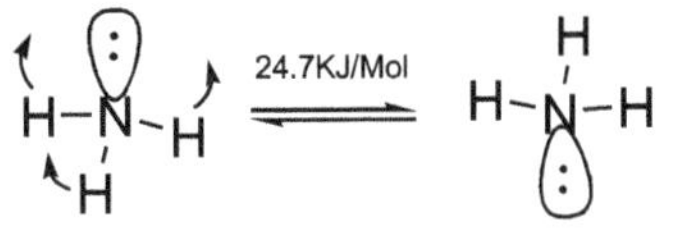

$$.....(1.29)$$

Ammonia in Biology

The role of ammonia in biology is immense. Ammonia plays a major part in maintaining the acid/base balance of normal animals. Diffusion of ammonia across renal tubules when combines with hydrogen ions, it allows excretion of excess acid. Ammonia is incorporated into biological molecules through glutamate and glutamine. The biosynthetic transformation of glutamine to glutamate is similar in all forms of life. Two reactions are involved in such transformation; firstly glutamate reacts with reacts with ATP to form a phosphate. The ATP gets converted to ADP by this process. The second step is the reaction of ammonium cation with the phosphates derivative to form glutamine. Both these reactions occur by the action of an enzyme called glutamine synthetase.

$$\text{Glutamate} + NH_4^+ + ATP \longrightarrow \text{Glutamine} + ADP + Pi + H^+$$

$$Pi = \text{phosphate anion} \qquad .....(1.30)$$

Ammonia is formed in biological systems from atmospheric nitrogen through certain enzymes. The dinitrogen gas itself is very stable and highly unreactive. Generally for the reduction of nitrogen to ammonia high pressure and temperature are required. But in biological system such reduction reactions take place at ambient condition. In biology the high activation barrier is partly overcome by binding to active site and facilitated by the energy dissipated during hydrolysis of adenosine triphosphate (equation 1.31).

$$N_2 + 10H^+ + 8e + 16ATP = 2NH_4^+ + 16ADP + 16\,Pi + H_2 \ (Pi \text{ is phosphate})$$

$$.....(1.31)$$

The process is carried out by an enzyme called nitrogenase; which has highly complex structure. The enzyme has two key components; one is called dinitrogenase reductase and other is dintrogenase. Dinitrogenase reductase contains tetranuclear iron Fe_4S_4 redox core. Its structural features are studied by preparing model complexes. The structure and reactivity correlation are achieved through understanding of their Mossbauer spectra and electron spin resonance spectra at different stages. Such studies are also supported by conventional spectroscopic study with UV-visible spectroscopy, infra-red spectroscopy, and cyclic voltametry etc. The core Fe_4S_4 can be oxidized or reduced by one electron. It has two binding sites for ATP. The nitrogenase contains iron and molybdenum ions. Its redox centers comprise of two molybdenum, thirty-two iron centers, and thirty sulphur centers per tetramer. Nitrogen fixation takes place at highly reduced form of dinitrogenase. The system provides the necessary six electrons for reduction of nitrogen and two electrons for the production of hydrogen. Generally the role of ATP in such reactions is to provide the energy required by transforming it to its corresponding diphosphate (ADP). But in this case it also helps in preorganising the enzyme to readily bind dinitrogen. The binding of ATP to enzyme and reorganization leads to lowering of the reduction potential of the protein which enhances the reducing capability.

Some reactions of borane

By virtue inherent Lewis acidity of borane, it forms adducts with different amines, ethers etc. through dative bonds. Some common adducts of borane are pyridine-borane, dimethylamine-borane, dimethylsulphide-borane, tetrahydrofuran-borane. The diborane reacts with two equivalent of ammonia to give $[BH_2(NH_3)_2]BH_4$. Such adducts are represented with $R_3N{\rightarrow}BX_3$ or $R_3N^+\text{-}BX_3^-$; suggests the origin of bonding electron from nitrogen in the dative bond. It does not mean that the nitrogen has a +ve charge and the boron has a negative. The co-ordination reduces the electron density of the donor but it is not sufficient to reverse the positive charge originally present on the uncoordinated boron atom of borane. This is reflected in the reactivity. Various nucleophiles react with these adducts at the boron centers, whereas electrophile reacts at the nitrogen centers. The diborane reacts with excess ammonia to give $B_3N_3H_6$ known as borazine.

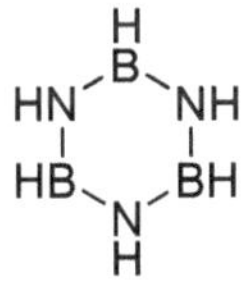

$$\begin{array}{c} \text{H} \\ | \\ \text{B} \\ \diagup \quad \diagdown \\ \text{HN} \qquad \text{NH} \\ | \qquad\qquad | \\ \text{HB} \qquad \text{BH} \\ \diagdown \quad \diagup \\ \text{N} \\ | \\ \text{H} \end{array}$$

Fig. 1.22 Structure of Borazine.

The metal borohydrides are formed by reactions of metal hydrides with boranes. For example, sodium borohydride is formed by reaction of borane with sodium hydride. Sodium borohydride is an excellent reducing agent for organic substrates.

$$BH_3 + NaH \longrightarrow NaBH_4 \qquad\qquad \dots(1.32)$$

The reaction of metal borohydride with iodine results in the formation of borane (equation 1.33). The borane is a good reducing agent as well as hydroborating agent. Borohydrides are also useful in the reduction of metal salts to metallic form. The metallic particles with nanometer diameters are prepared reactions of metal salts with metal borohydrides.

$$NaBH_4 + I_2 \longrightarrow BH_3 + NaI \qquad\qquad .....(1.33)$$

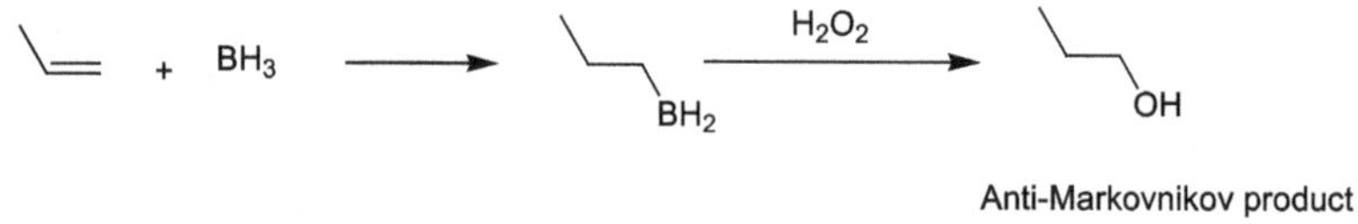

Scheme 1.17 Hydroboration reaction.

Further to these, the boranes and their various amine adducts are identified as storage materials for hydrogen gas and are prospective candidates for fuel cells.

Remarks

The correlation of experimental results or theoretical prediction to design new materials with new properties arises from need of studying a particular problem. A general observation may arise from series of experiments or some phenomenon occurring in nature. Extensive experimentation is needed to formulate a general law to make it widely applicable. Any theoretical model needs to be tested with experimental data for their validity. Understanding of the exceptions from extrapolation of a theory makes the theory more acceptable. Deviation from an ideal system is a common feature of experimental chemistry. For example, the relation of activity coefficient with ionic strength for strong electrolyte is given by Debye theory. While in practice, many experimental data do not tally to the predicted value. In such cases additional parameters are to be added. Nevertheless, the adopted theory in this case can be a basis for fundamental understanding. Next four chapters of this book is on some essential experiments for basic understanding of chemistry, chemical biology, pharmacology, *with the pertinent* principles behind them.

CHAPTER 2

Basis of Experimental Chemistry

General Chemistry

Some basic principles

Chemical reactions are associated with chemical compounds as reactants, reagents, intermediates, and products. They are also associated with the parameters such as (a) reaction conditions and (b) reaction coordinates involving variables like time, concentration etc. In physical chemistry, the last two parameters are dealt in more depth. The reaction conditions have references to the external and internal conditions maintained while performing a chemical reaction. It may be pressure, temperature, concentration, pH, ionic strength, and external energy supplied in the form of heat, light, or electricity. The term reaction coordinate, describes the parameters that are needed to understand the progress of a chemical reaction. Such parameters may be reaction time, concentration, change of any other physical parameters such as pressure, temperature, voltage etc., that varies with the course of reactions. The graphical representations of the variation of these parameters during the course of a reaction are called the reaction profile. In a sense these profiles can explain the course of reactions.

To understand the different processes occurring while carrying out any chemistry experiment or as a whole experiments of other subjects, it is essential to identify a directly or indirectly measurable parameter that will be capable of describing the process under consideration. In chemistry, such parameters may be any physical or thermodynamic quantities. Thus, for executing any experiment in chemistry, knowledge on physical quantities and also knowledge to determine certain physical quantities associated with a particular experiment is essential. To study chemistry of the experiments that are performed in solution, the first point comes to mind is the solubility of a compound in a particular solvent. For this purpose, a prior knowledge on polarity of the solvent and idea about the nature of the solute are essential. For example, an ionic compound will be more soluble in protic or polar solvents; whereas, covalent compounds are more soluble in nonpolar or less polar solvents. The polarity of the solvent is ascertained from dielectric constant of the solvent. There are number of ways to express solubility of a compound (solute) in a solvent. For example, it may be in terms of weight by weight, weight by volume, volume by volume of a solute dissolved in a solvent. At times, for these purposes conversion of weight to volume becomes necessary for a liquid, thus, the density of a liquid is to be determined at a particular temperature. The densities of liquids are easily determined by using specific gravity bottle. In such method, density is determined by weighing one ml of water as well as the liquid independently. The density of a solvent or solute (liquid) is calculated from the ratio of these two weights at a particular temperature. However, if such measurements are not carried out at standard temperature and pressure, corrections to these values are to be made by using the density of water from available literature value as reference.

Over and above the simple ways of preparation of solution by methods such as weight by volume, volume by volume, the concentration of solutions are more precisely expressed in terms of molarity, molality, normality, and mole fraction. Each of these scales has their special interest; as a particular scale is appropriate and useful for determination of certain specific physical properties. Expressing concentration in term of molarity is useful in calculating reaction yield and it is also useful for describing stoichiometry of a chemical reaction. The term normality is associated with electron transfer processes; and it also relates the numbers of replaceable hydrogen atoms; so normality scale is very useful in understanding redox reactions and also in acid base chemistry. This scale is especially in great uses in quantitative estimations. Molality is useful in determination of colligative properties, while mole fraction is useful in understanding thermodynamic properties associated with partial molar quantities.

Solubility

Preparation of a solution of exact strength becomes essential in quantitative estimation experiments or to maintain correct stoichiometry while performing a reaction; or in measuring a physical quantity. The dilute solutions with exact concentration are generally prepared by dilution method.

To do so, a standard solution is prepared; the strength of such solution is re-determined by appropriate titration. The standard solution thus prepared, is diluted to correct volume to make another solution of exact strength. Thus, for making any solution in a particular solvent the solubility is of prime concern. A compound is soluble in a solvent until a point is reached; it is called saturation of a solution. The solubility of a solute in solvent depends on temperature. The determination of solubility of a solid solute in a solvent can be carried out in a tube containing solvent with a stirring facility. For this purpose solubility is measured by preparing saturated solution by slow addition of solute and observing the dissolution visually. Such measurements may be inaccurate; as it is purely dependent on the skill of the experimentalist. There are many other ways; such as refractive index measurement, conductance measurement, electromotive force measurement, and spectroscopic analysis to accurately determine solubility of a solute. The solubility of compounds is also studied by titrating a saturated solution with appropriate titrant.

Solubility by refractive index measurement

The solubility of a compound in a particular solvent can be determined by measurement of refractive index. For this purpose, refractive indices of a series of known samples are measured and a calibration graph is made. For a particular unknown solution the refractive index is determined. The refractive index thus measured, can be fitted in appropriate place of the calibration graph, and from the plot one can determine the solubility of the compound. Consider a solution comprising of two components of liquid substrates. Let n_1 and n_2 be the refractive indices of two components; namely solute and solvent. Let the solute and solvent has the density d_1 and d_2 respectively. Let n_3 and d_3 to be the refractive index and density of the solution under consideration. The percentage x amount of the one component in the solution can be calculated from the expression:

$$(n_1 - 1)\, x/d_1 = [100\, (n_3 - 1)/d_3] - [(n_2 - 1)\, (100 - x)/d_2] \qquad .....(2.1)$$

Solubility by conductance measurement

The solubility can be also determined from conductance measurements. For this purpose the compound should be sparingly soluble in the particular solvent and should not get hydrolysed in the solvent. Let the limiting conductance offered by such a solution is Λo. This conductance value is the sum of ionic conductance of each ion present in the solution. The measurement of equivalent conductance provides information about the concentration of the solution of sparingly soluble salt. The solubility of a sparingly soluble salt is calculated by

$$C = 1000\, \kappa / \Lambda o.$$

where, κ is the specific conductance.

Solubility by electromotive force measurement

Let us consider a sparingly soluble salt MX to determine its solubility by electromotive force (e.m.f) measurement. Since the compound is sparingly soluble; it offers common ion effect. This happens, as there is excess insoluble and undissociated salt MX present in the system. The ionization of MX is given by

$$MX \rightleftharpoons M^+ + X^- \qquad \qquad .....(2.2)$$

$K_{sp} = [M^+][X^-]$. As the MX is in excess; the [MX] is constant. K_{sp} is the solubility product.

For MX in pure water $[M^+] = [X^-]$. Hence, solubility in this particular case is the square root of K_{sp}. Although, these expressions are shown in terms of concentrations; for accurate determination of solubility the activity of these species are to be considered.

Let us consider an electrochemical cell

$$Ag \mid AgCl,\ KCl\ (0.01M) \mid KNO_3\ sat \mid AgNO_3\ (0.01M) \mid Ag \qquad .....(2.3)$$

This cell is suitable for determination of solubility of silver chloride in water. The e.m.f of the cell without a junction is governed by

$$E = 0.05916 \log \{[aAg^+_s]/[aAg^+_{sol}]\}.$$

where aAg^+_{sol} is the activity of silver ion in 0.01 M potassium chloride solution, and the aAg^+_s is the activity of silver ion in 0.01M silver nitrate solution.

These quantities can be determined by constructing two separate cells connected by a salt bridge. By measuring the electromotive force of this cell, the concentration of the $[M^+]$ ion in solution can be determined.

Hydrogen ion concentration

Measurement of acidity of a solution is very essential in understanding chemical reactivity of a substrate; it is also useful to maintain pH of a solution. The acidity of a protic acid can be determined by finding out the equivalent point by titration of the acid with a strong base. However, the methods based on pH and conductance measurements have advantages to determine the end point of an acid base titration. These methods provide means to accurately determine the equivalent point by graphical method. These kinds of measurements are also useful in understanding steps of dissociation of a poly-basic acid or a poly-acidic base that can have multiple dissociation steps. The pH measurements during an acid base titration can be also used to determine the pK_a of a weak acid. Consider the dissociation of acetic acid, in water. It dissociates as

$$CH_3COOH + H_2O \ = \ CH_3COO^- + H_3O^+ \qquad .....(2.4)$$

Since acetic acid is a weak acid, a fraction of it only dissociates in solution. Thus,

$$K_a = [H_3O^+]\,[CH_3COO^-]\,/\,[CH_3COOH] \qquad .....(2.5)$$

$$pH = pK_a - \log\,[CH_3COOH]\,/\,[CH_3COO^-] \qquad .....(2.6)$$

where,

$$pK_a = -\log[K_a]$$

The equation 2.6 is known as Henderson's equation. When acetic acid is titrated with a strong base, there will be a point during the titration at which the number of equivalents of base is half of the number of the equivalents of the acetic acid present. This is the point at which 50% of the acid has been titrated to produce acetate anion and 50% remains as acetic acid. At this half - equivalence point, $[CH_3COOH] = [CH_3COO^-]$ and

$$pH = pK_a \qquad .....(2.7)$$

Recording the pH versus the volume of base added, while titrating a weak acid against a strong base, the dissociation constant of the weak acid can be obtained. The graph obtained from changes in pH vs volume of alkali added from such titration is shown in Fig. 2.1. From the graph the strength of the acid as well as the pK_a of the acid can be determined.

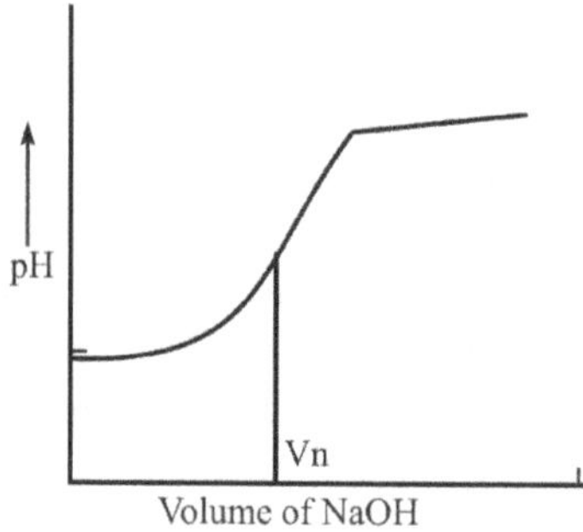

Fig. 2.1 pH-Titration of a weak acid with strong base.

The equivalence point / neutralisation point of a titration can also be determined by measuring the conductivity of a solution during the course of titration. The plot of conductance vs amount of a reactant added as titrant in different titrations varies, the types of plots depend on the nature of the acid, base or titrant used for the particular titration. This is due to the reason that electrical conductance of a solution depends upon the number of ions present in the solution. The total conductance offered by a solution is the summation of conductivity of the ions present in the solution and the conductance of individual ions depends on the ionic mobilities. Thus, if a strong acid is titrated with a strong base the conductance curve of the titration will be different from that of the titration of a weak acid and a weak base. Thus, for a titration of a strong acid with a strong base, the plot of conductance *versus* the volume of titrant added, two straight lines with different slopes are obtained. The point of intersection is the equivalence point. Consider the titration of a strong acid like hydrochloric acid, with a strong base, like sodium hydroxide.

$$[H^+ + Cl^-] + [Na^+ + OH^-] \Rightarrow [Na^+ + Cl^-] + H_2O \qquad \text{.....(2.8)}$$

Before the base is added the acid solution has a high conductance due to the highly mobile hydrogen ions. When sodium hydroxide is added to the strong acid the highly mobile hydrogen ions are replaced by less mobile Na^+ ions, thereby, decreasing the conductance of the solution. When the neutralisation process is completed, further addition of alkali will cause the conductance to increase, owing to the excess of highly mobile OH^- ions. Thus, the conductance will have a minimum value at the equivalence point.

A plot of conductance versus the volume of titrant consists of two straight lines intersecting at the neutralisation point.

In this experiment conductance is plotted as arbitrary units in the Y-axis is not calibrated. However, the conductance needs to be calibrated if the conductance values are to be used for other purposes. In such cases it is appropriate to determine the specific conductance, equivalent conductance or molar conductance of the solution. For the measurement of these quantities the cell constant is also to be determined. The cell constant is determined by measuring the conductance offered by standard potassium chloride solution. The ratio between the actual conductance offered by the cell under use to the standard conductance value obtained from cell having unit cell constant is the cell constant of the cell under consideration.

Let Λ be the conductance offered by a solution, and κ be the specific conductance of the solution; then we have:

Conductance = Specific conductance / cell constant

$$\Lambda = \kappa / \frac{1}{a} \qquad \qquad .....(2.9)$$

l = Distance between the electrodes

a = Area of each electrode

The factor l/a is a constant for a particular cell and is known as cell constant (z)

$$\Lambda = \kappa / z$$

Values of the specific conductance of the solutions of potassium chloride at different concentration and at different temperatures can be known from standard tables that are available in literature. Knowing the cell constant of the cell, the specific conductance of a solution can be found out. Knowing the specific conductance of a solution, its equivalent conductance (Λ_c) can be calculated from the equation 2.10:

$$\Lambda_c = \frac{1000\ k}{C} \qquad \qquad .....(2.10)$$

where C is the concentration of the solution expressed in normality i.e. in gm equivalent per litre.

Buffer solutions

Maintaining a constant pH throughout a chemical reaction or to make a measurement at a particular hydrogen ion concentration, pH is very often needed in biology as well as in chemistry. Addition of certain solutions to another solution helps in maintaining constant pH of a solution on slight increase or decrease of hydrogen or hydroxyl ion concentration. Such solution that is added to maintain constant pH of another solution is called a Buffer solution. Buffer solutions are needed to adjust pH of enzymes in many organisms to work at ordinary condition. Blood plasma contains a buffer of carbonic acid and bicarbonate, which maintains a pH between 7.35 and 7.45. Buffer solutions are used in fermentation industry and for setting color of dyes used in fabrics. For these purposes, buffer solutions are prepared in laboratory by choosing appropriate combinations of acid and salt; or base and salt. Buffer tablets of different pH are commercially available. The principle of Buffer action is based on common ion effect. To generate buffer of different pH, different combinations of chemicals are used. For example, consider a mixture of acetic acid and sodium acetate; acetic acid ionizes as

$$CH_3COOH \rightleftharpoons CH_3COO^- + H^+$$

$$.....(2.11)$$

Addition of sodium acetate to the solution pushes the equilibrium towards left. Addition of hydroxide from external source removes the H^+ and the acetate prevents this. Thus, no change in hydrogen ion concentration takes place when a small amount of base is added.

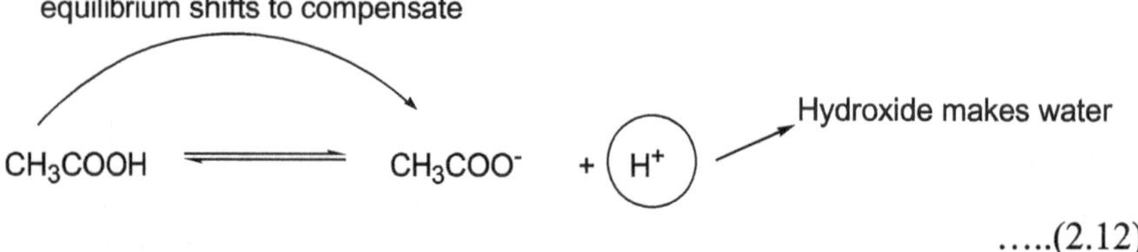

$$.....(2.12)$$

Ammonia dissolved in water forms ammonium hydroxide. Ammonium hydroxide ionizes as

$$NH_3 + H_2O \longrightarrow NH_4OH \rightleftharpoons NH_4^+ + OH^- \quad .....(2.13)$$

Presence of ammonium acetate in such solution will push the equilibrium towards left. Addition of hydrogen ions to this ionization equilibrium will produce water. As soon as this happens, the equilibrium ammonium ion takes over the equilibrium. This keeps on happening until most of the hydrogen ions are removed (equation 2.14).

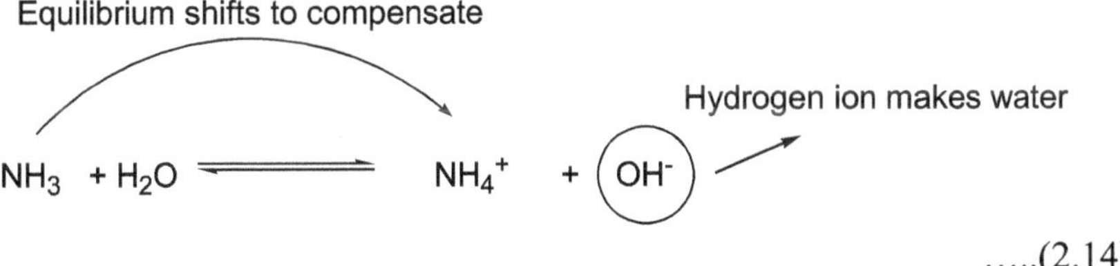

$$.....(2.14)$$

Buffer capacity is a quantitative measure of the resistance of a buffer solution to pH change on addition of hydroxide ions.

Mathematically buffer capacity is

$$\beta = dn/dpH \qquad .... (2.15)$$

where n is the number equivalent of a base added to the solution. This equation defines change of pH with base concentration, however addition of an acid will also do the same; but exactly in opposite direction i.e. a negative sign will be enough to describe the action of acid. When a non-protic, strong base is added to buffer solution, the strong base will get protonated. It will cause a shift in the equilibrium. However, the charge neutrality of the solution is to be maintained. Thus,

$$[A^-] + [OH^-] = [B^+] + [H^+] \qquad .....(2.16)$$

where, B^+ is the protonated strong base added. Total concentration of the buffer will C_{buff} will be

$$C_{buff} = [HA] + [A^-] \qquad .....(2.17)$$

with dissociation:

$$[HA] = [H^+][A^-]/K_a \qquad .....(2.18)$$

and

$$C_{buff} = [H^+][A^-]/K_a + [A^-] \qquad .....(2.19)$$

$$[A^-] = C_{buff} K_a / (K_a + [H^+]) \qquad .....(2.20)$$

From the above equations

$$n = K_w/[H^+] - [H^+] + C_{buff} K_a/ (K_a + [H^+]) \qquad .....(2.21)$$

Differentiating with respect to pH

$$\beta = dn/dpH = dn/d[H^+]\, d[H^+]/dpH$$

$$\beta = \{- Kw/[H^+]^2 - 1 - C_{buff}\, K_a/\, (K_a + [H^+])^2\,\} \, (-2.303\,[H^+])$$

$$.....(2.22)$$

or

$$\beta = 2.303\, (K_w/[H^+] + [H^+] + C_{buff}\, K_a\, [H^+]/\, (K_a + [H^+]^2) \qquad .....(2.23)$$

From the equation 2.23 one can ascertain the range of pH at which buffer capacity of a buffer will be highest. In general, a buffer solution is prepared from a weak acid and its conjugate base; such buffers generally work in a narrow pH range. When a wide range of pH is needed, a buffer capable of working in a wider range of pH can be prepared by mixing the two buffering agents. Different composition of buffer works in different pH ranges; for example, hydrochloric acid-sodium citrate works at pH, 1-5; citric acid, sodium citrate works in the range 2.5-5.6; acetic acid - sodium acetate buffer works at 3.7-5.6; sodium dihydrogen phosphate-sodium hydrogen phosphate works in the range of 6-9; borax and sodium hydroxide buffer works in the pH range of 9.2-11.

Ionic strength

The ionic strength is an important parameter defined as:

$$I = \tfrac{1}{2} \Sigma\, C_i z_i^2 \qquad\qquad .....(2.24)$$

where C_i is a molar concentration of ith ion present in the solution and z_i is its charge. Summation is done for all charged molecules present in the solution.

Sea water has ionic strength of about 0.7 and many common laboratory reagents have much higher ionic strength, thus the reactions carried out or experiments carried with such solvents or reactions carried out by reagents may be misleading unless the effect of ionic strength are accounted. In fact, the ionic strength affects the activity coefficient. The ionic strength is related to activity coefficient by:

$$\log f_z = -0.51\, Z^2\, \sqrt{I}\, /\, (I + \sqrt{I}) \qquad .....(2.25a)$$

Or, more approximately,

$$\log f_z = A.\, Z^+\, Z^-\, \sqrt{I} \qquad\qquad .....(2.25b)$$

Activity of an ion is related to concentration by $a_{ion} = f_z C_i$, where f_z is the activity coefficient. Thus, for studying any equilibrium in a solution the activity is to be taken instead of concentration. Inclusion of activity coefficient with concentration term enables one to understand the related thermodynamic parameters under non-ideal conditions.

Solubility vs ionic strength

Solubility product of a compound can be determined by measuring the ionic strengths of solutions. The solubility product is defined as the product of the activities of the ions. For example, consider a sparingly soluble salt namely lead iodide that ionizes as equation 2.26.

$$PbI_2 = Pb^{2+} + 2I^- \qquad\qquad .....(2.26)$$

$$K_a = a_{pb}^{2+} (a_I^-)^2 = C_{Pb}^{2+} (C_I^-)^2 (f_\pm)^3 = K_C (f_\pm)^3 \qquad .....(2.27)$$

where, $f_\pm$ is the mean activity coefficient of the electrolyte.

or,

$$\log Kc = \log K_a - 3 \log (f_\pm)$$
$$= \log K_a + 3 \, A.Z^+. \, Z^-.I^{1/2} \qquad\qquad .....(2.28)$$

(by using equation 2.25b)

where

$$K_c = (C_{Pb}^{2+}) (C_I^-)^2 \qquad\qquad .....(2.29)$$

A plot of $\log K_c$ vs. $I^{1/2}$ gives a straight line with intercept equal to $\log K_a$ and hence K_a can be determined.

Determination of K_a of lead(II) nitrate

Prepare a series of solutions lead(II) nitrate ranging from 0.25 M to 0.6 M in water. Take each lead(II) nitrate solution (10 ml) and titrate with potassium iodide (0.05 M) solution independently. In each case, at the end point the solution turns yellow. Yellow precipitate shows the saturation of solution. Calculate the final concentration of lead nitrate and the potassium iodide in the total volume of the solution in each case. Further, calculate concentration solubility product and ionic strength of each of the solutions by using equation 2.29 and 2.24. Plot $\log K_c$ vs. $I^{1/2}$ to get the value of K_a.

Esterification of an acid

Acid base chemistry finds important applications in chemical reactions, catalysis, stabilisation of compounds, manufacturing of industrially important compounds. The acid base chemistry is also used for synthesis of various compounds of medicinal value. One of the simple synthetic reactions based on such chemistry is the acid catalysed esterification reaction of an acid. Consider the esterification of benzoic acid with methanol in the presence of an acid catalyst; as illustrated in equation 2.30:

$$\text{PhCOOH} + CH_3OH \underset{}{\overset{acid}{\rightleftharpoons}} \text{PhCOOMe} + H_2O$$

.....(2.30)

Such esterification reactions are useful in making medicines, for example, methyl salicylate is a pain relieving substance, which is the methyl ester of salicylic acid. Esters generally have sweet smells, so they are also useful for making perfumes.

Preparation of methyl benzoate

Take benzoic acid (3.05 g 0.025 mol), methanol (6 ml, 0.15mole), along with concentrated sulphuric acid (0.5 ml) in a round-bottomed flask. Fit the flask with a reflux condenser and boil the mixture for about 45 minutes by placing it in an oil bath. Cool the reaction mixture to room temperature, add 30 ml ether. Transfer the mixture into a separatory funnel containing 15 ml cold water. Wash the ether layer with 10 ml ice cold water and 10 ml 5% aqueous sodium carbonate solution. Dry the ethereal extract over anhydrous magnesium sulphate or anhydrous sodium sulphate. Evaporate the ether to obtain the methyl benzoate as transparent oily lquid.

Soap

Soaps are salts of fatty acids. The fatty acids are consists of long chain hydrocarbon attached to carboxylic acid group at one end. The carboxylate groups are generally associated with an alkali metal ion, sodium or potassium ions and are in the form of salts. The structure of a soap molecule has composition $CH_3–(CH_2)_n–COO^-Na^+$ where n is an integer; which generally is greater than 10. The base hydrolysis of fats and oils is used for synthesis of soap. The IR spectra of soap vary from source to source. But the characteristic peaks of carboxylic acid functional groups are present in all cases. The IR spectrum of a soap prepared from vegetable ghee is shown in Fig. 2.2. It has characteristic ν_{OH} at 3448 cm^{-1}, ν_{C-H} at 2928 cm^{-1}, overtone of ν_{OH} at 1743 cm^{-1}. The carboxylate group stretching appears at 1634 cm^{-1}.

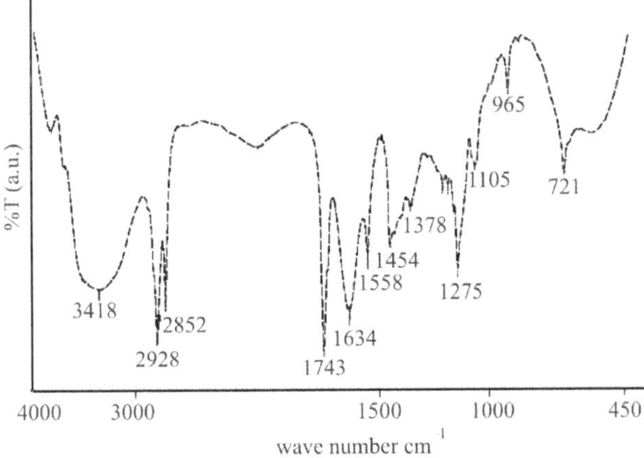

Fig. 2.2 FT-IR spectra (KBr, cm^{-1}) of soap from vegetable ghee.

Preparation of soap

Dissolve sodium hydroxide (2 g), in a mixed solvent of ethanol (10 ml) and distilled water (10 ml) in a beaker (100 ml) by continuous stirring. Place vegetable ghee (2 g) in another flask (125 ml) and add the above mentioned sodium hydroxide solution to it. Heat the solution over a water bath and allow the reaction to proceed at least for 30 minutes. In the mean time, prepare a solution of sodium chloride by dissolving about 15 g of sodium chloride in 80 ml of water in a beaker (500 ml). Add about 25 g ice to this solution. After 30 minutes of heating the reaction mixture, pour the hot reaction mixture into this cold salt solution. Stir the resulting mixture for about five minutes. Decant the water and collect the paste, which is the desired soap; wash the collected soap twice with 10 ml portions of ice-cold water. Dry the soap by pressing it over tissue paper.

Alkalinity of soap

The alkalinity of a soap solution can be tested with the help of a red litmus paper or by dissolving soap (0.05 g) in hot water and adding a drop of phenolphthalein indicator. The solution turns red showing its alkaline behavior.

Base catalysed phase transfer reactions

The understanding of acid-base reactions also helps in developing catalytic cycles for phase transfer catalysis. For example, the reaction of cyclohexene with chloroform in strong alkaline medium is catalysed by tetrabutyl-ammonium chloride (equation 2.31).

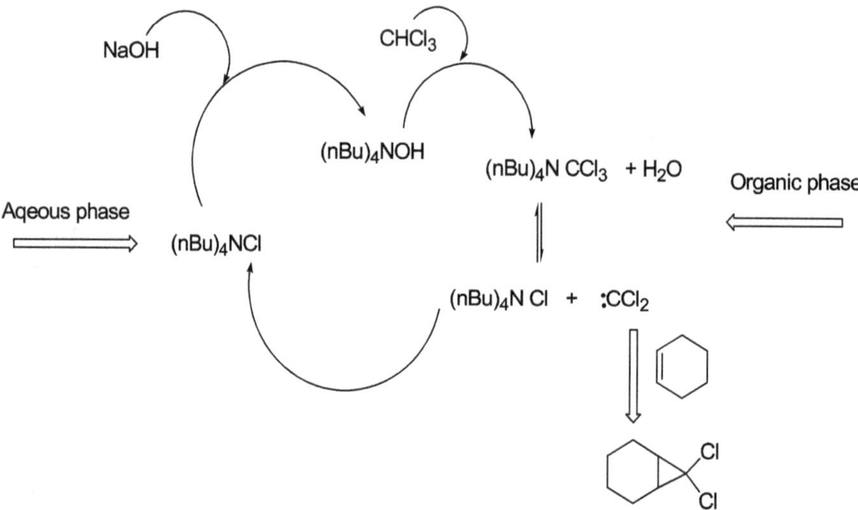

.....(2.31)

This reaction can be carried out under phase transfer condition with chloroform and water as mixed solvent. In this reaction chloroform acts as a solvent as well as a reactant. The catalytic cycle is shown in the scheme 2.1.

Scheme 2.1 Phase transfer catalysis based on acid base reactions.

The key step in this reaction is the abstraction of proton from chloroform by tetrabutylammonium hydroxide, which acts as phase transfer reagent; it also assists in generation of dichlorocarbene. The tetrabutylammonium chloride is formed in the reaction, dissolves in organic phase. This compound is brought to the aqueous phase by reacting with sodium hydroxide to give tetrabutylammonium hydroxide.

Decarboxylation reaction by base

Basic hydrolysis of an ester may lead to decarboxylation of a β-ketoester. For example, methyl acetoacetate on reaction with potassium hydroxide gives corresponding β-ketoacid, namely, acetoacetic acid. The acetoacetic acid is unstable in basic medium and loses carbon dioxide to give acetone (scheme 2.2).

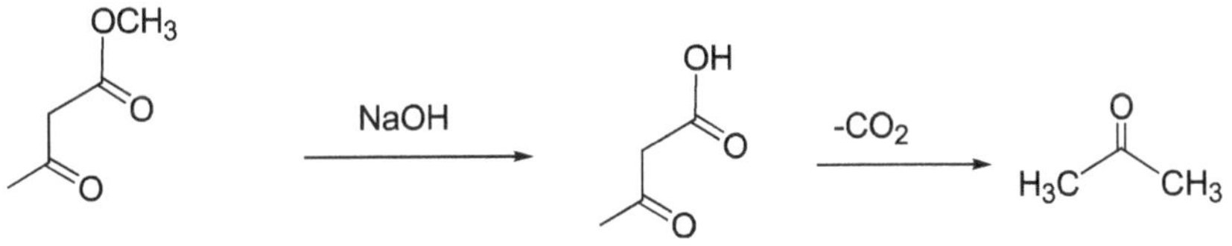

Scheme 2.2 Hydrolysis and decarboxylation.

This reaction can be studied by boiling an alkaline solution of methyl acetoacetate; followed by converting the acetone formed in the reaction to its phenyl hydrazone derivative. The reaction can also be monitored by gas chromatography by checking the products formed as time progresses.

Hydrolysis equilibrium

Let us consider the protonation equilibrium of an amine

$$RNH_2 \; + \; H_2O \; \rightleftharpoons RNH_3^+ \; + \; {}^-OH \qquad \qquad .....(2.32)$$

Each amine is associated with a hydrolytic equilibrium constant. Depending on the magnitude of such equilibrium constants, a comparison on the ease of hydrolysis among series of amines can be studied. For example, the acidity difference of aniline in comparison to that of the cyclohexylamine, can be explained from this equilibrium. Aniline has one lone pair of electrons on the nitrogen, which is resonance stabilized. Whereas the anilinium ion has an isolated aromatic system without the involvement of the $^+NH_3$ group in resonance structure; thus the later has less stability than aniline. For such reasons, protonation of aniline is as difficult as that of the protonation of cyclohexylamine, which has no conjugation effect in the parent compound as well as in the protonated form. Similar logic is valid for weak basicity of pyrole; it is a weak base as its protonated species loses the aromaticity and is less stable than the parent compound (equation 2.33).

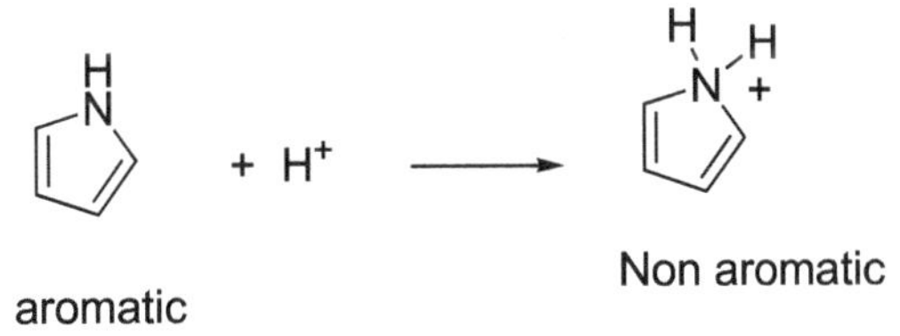

aromatic

Non aromatic

$$.....(2.33)$$

Ammonium phosphate

Ammonium phosphate is a popular and effective fertilizer. The reaction of ammonium hydroxide and phosphoric acid is used to prepare ammonium phosphate fertilizers.

$$3\ NH_4OH + H_3PO_4 \rightarrow (NH_4)_3\ PO_4 + 3\ H_2O \qquad \dots\dots(2.34)$$

Method of preparation

Take 20 ml phosphoric acid in a beaker (100 ml). Add ammonia solution to the phosphoric acid in portions at a time with constant stirring until the pH of the reaction mixture reaches 7 or slightly above; test this with a pH indicator paper. Evaporate the solution to one fifth of its original volume (care should be taken while heating; so that the solution do not bump out of the reaction vessel) and cool. On cooling, a white crystalline solid is formed. Filter off the crystals and dry. This solid is ammonium phosphate.

Estimation of amount of phosphate in ammonium phosphate

Take ammonium phosphate (0.2 g, 1.3 mmol) in water (10 ml). Add 1 drop of methyl red indicator followed by ammonia buffer (10 ml, pH = 10). Yellow colour appears. To this solution add magnesium sulphate heptahydrate (0.98 gm, 4.0 mmol) and stir. Allow the solution to stand for one hour. Filter the solution to a conical flask (100 ml) and take the filtrate. To the filtrate add buffer solution to make the pH = 10. Titrate the solution with a standard ethylenediamine tetra-acetic acid (0.05 M) solution using erichrome black-T as an indicator. The end point will be seen upon formation of blue coloration of solution. Thus, phosphate is estimated by knowing the amount of magnesium that reacted with it to precipitate as magnesium phosphate.

Potash alum

Potash alum is one of the most widely used industrial chemical. It is mainly due its use in the pulp and paper industry and in water treatment plants. It is effective for a broad range of water treatment problems; it can function as a coagulant, flocculent, precipitant and emulsion breaker. Alum removes turbidity, suspended solids, colloids, and colors from water. Alum is prepared by acid base chemistry by reacting aluminium with potassium hydroxide. The reaction gives potassium aluminate; which on treatment with sulphuric acid gives potash alum.

$$2Al + 2\ KOH + 6H_2O \longrightarrow 2\ KAl(OH)_4 + 3\ H_2O \qquad \dots\dots(2.35)$$

$$2\ KAl(OH)_4 + 4\ H_2SO_4 \longrightarrow K_2SO_4\ Al_2SO_4.\ 24\ H_2O \qquad \dots\dots(2.36)$$

Method of preparation of potash alum

Take finely cut aluminum scrap (0.5 gm) and place them in a beaker (250 ml). Add a solution of potassium hydroxide solution (1.75 gm in 15 ml water) carefully to the aluminium scrap. Heat the mixture in the beaker gently till all the aluminum dissolves. Care should be taken as a good amount of effervescence takes place. An ash-colored solution will be formed. Filter the warm solution carefully through cotton or glass wool. Cool the solution; and slowly add sulphuric acid (15 ml, 6 M) to the solution with constant stirring. The solution will give some solid precipitate on addition of acid. Heat the solution to dissolve the solid. If the solution is still not clear, filter the mixture again while it is warm. Cool the clear solution in an ice-bath for 20 mins to get crystals of alum. Recrystallize the product from water.

Role of acid and base in carbonyl activation

The role of an acid catalyst in a chemical reaction is relatively easy to understand; but they have tremendous implications in organic chemistry and biology. Whether it is the acidity caused in the intestine due to indigestion or the color of red rose, the acid concentration is the key factor in these processes. Acid, base chemistry comes into picture while generation of cations, from C=C double bonded compounds, or in the reactions of carbonyl compounds and so on. Let us consider the protonation of the oxygen atom of a carbonyl group; the protonation increases the polarity of carbonyl group. Such protonations affect the rate of reaction of a nucleophile at the carbonyl center. The situation is described below in the equilibrium 2.37,

$$\cdots\cdots(2.37)$$

As the protonation takes place the carbon of the carbonyl group gets polarized to make a positive center at the carbon of the carbonyl group. This being the first step in an acid catalysed hydrolytic attack on a carbonyl compound, the protonation plays a decisive role in the rate and selectivity:

$$\cdots\cdots(2.38)$$

Loss of a proton from such protonated species leads to formation of diols as shown below:

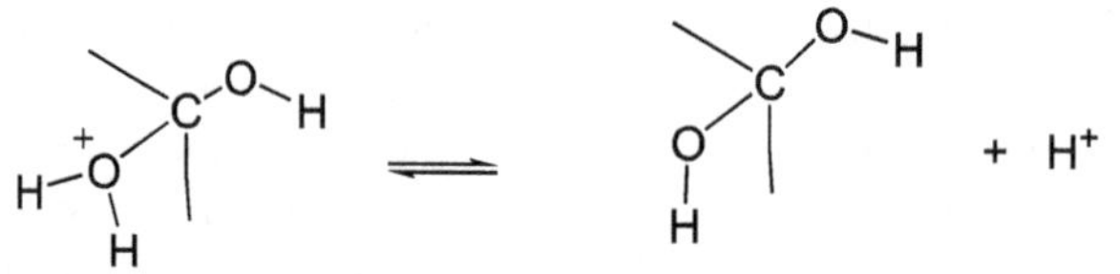

$$\ldots\ldots(2.39)$$

The OH⁻ can also cause similar equilibrium. The OH⁻ being a strong nucleophile can attack of its own. It involves following steps:

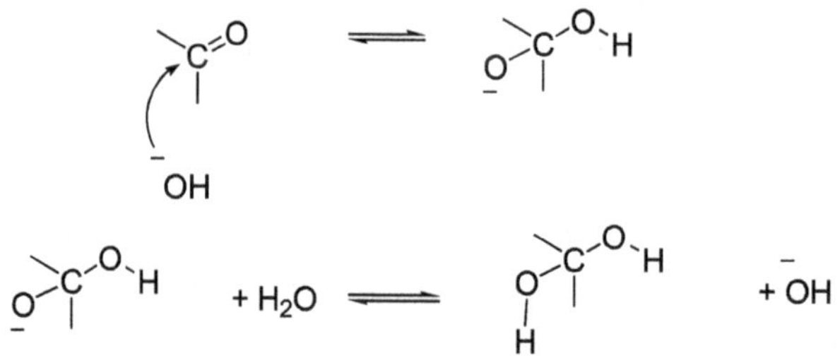

Scheme 3 Diol formation from carbonyl

From the above reaction schemes it is clear that a nucleophile or Lewis base attacks the positive end of the carbonyl group whereas a Lewis acid or electrophile at the other end of the carbonyl.

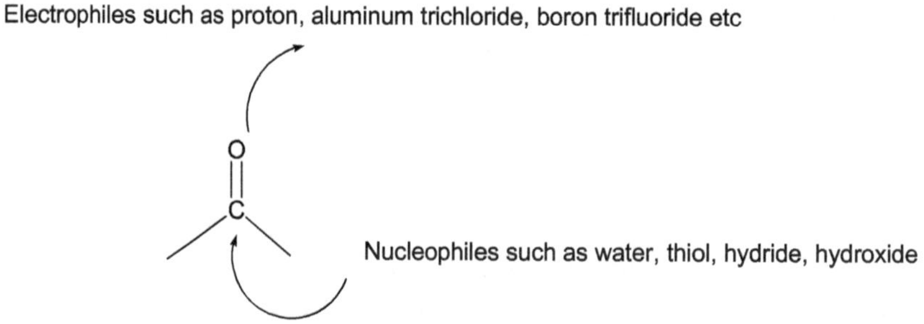

$$\ldots\ldots(2.40)$$

However, the presence of hydrogen atom/s on the α-carbon i.e. carbon next to carbonyl group can change the entire picture of such reactions. In such case, the base abstracts the hydrogen attached to the carbon at α-position and leads to formation of resonance stabilized carbonanion.

Hydrolysis of methyl acetate

To substantiate acid catalysed reactions of carbonyl compounds further; let us consider the reaction of hydrolysis of methyl acetate. The methyl acetate is hydrolyzed by a mineral acid to form methyl alcohol and acetic acid. The reaction is shown in equation 2.41.

$$CH_3COOCH_3 + H_2O \xrightarrow{H^+} CH_3COOH + CH_3OH \qquad .....(2.41)$$

In a dilute aqueous solution of the ester, concentration of water is in excess. It practically remains constant during the reaction. The concentration of H^+ ion which catalyses the hydrolysis reaction also remains constant. Thus, the rate of the reaction depends only on the concentration of the ester; and is represented by a pseudo first order kinetics as follows:

$$K = \frac{2.302}{t} \log \frac{a}{a - x} \qquad .....(2.42)$$

where k is the rate constant, a is the initial concentration of methyl acetate and x is the amount of methyl acetate hydrolyzed at time t.

The acetic acid is produced in this hydrolysis reaction. The kinetics of this reaction is followed by withdrawing a fixed volume of the reaction mixture from time to time and titrating them with standard sodium hydroxide solution. Let the volume of sodium hydroxide needed to titrate the reaction mixture at t = 0 is V_0;   this amount is equivalent to the initial amount of mineral acid used in the experiment. If the volume of sodium hydroxide consumed at time t and at the time of completion of the hydrolysis reaction are V_t and V_α respectively, then

$$a \; \alpha \; (V\alpha - V_0) \text{ and } (a-x) \; \alpha \; (V\alpha - Vt)$$

Thus,

$$k = \frac{2.302}{t} \log \frac{(V\alpha - V_0)}{(V\alpha - Vt)} \qquad .....(2.43)$$

Plot of log 1/ $(V_0 - V_t)$ vs time will be a straight line. The rate constant 'k' is calculated from the slope of the line. By carrying out this reaction at two different temperatures; two different rate constants will be obtained. From these rate constants, the activation energy for the reaction can be calculated.

Procedure for kinetics

Take hydrochloric acid (100 ml, 0.1M) in a dry stoppered conical flask (100 ml) and add distilled methyl acetate (10 ml). Record the temperature of the solution. Shake the solution to ensure proper mixing. Immediately after mixing, withdraw 5 ml of the reaction mixture into a conical flask containing ice-cold water (about 20 g of ice in 10 ml water) so as to arrest the reaction. Start the timer, when the solution from the pipette has been half discharged into the flask. Titrate this solution quickly with sodium hydroxide solution (0.05M) using phenolphthalein indicator (few drops, 0.5% aqueous). Formation of a faint pink color is the indication for the end point. This corresponds to V_0. Perform similar titrations by taking 5 ml reaction mixture from the reaction solution at successive time intervals of 5, 10, 15, 20 mins. These values correspond to V_t, at different times.

Take 25 ml of the reaction mixture into a separate stoppered conical flask; and heat the solution at about 50 ^{0}C for one hour to complete the hydrolysis. Cool the solution to room temperature. Take out 5 ml solution from this; and titrate with sodium hydroxide solution (0.05M) using one or two drops of phenolphthalein indicator. The titre value obtained is Vα. Calculate the rate constant from a graph of log $1/(V_0 - V_t)$ vs time.

Saponification of an ester

The base hydrolysis of an ester when applied to fats and oils is called saponification. These types of reactions also find important place in carbon-heteroatom and carbon-carbon bond formation reactions. Consider the hydrolysis reaction of ethylacetate in the presence of a base as shown below

$$CH_3COOC_2H_5 + NaOH \quad \rightarrow \quad CH_3COONa + C_2H_5OH \quad .....(2.44)$$

This reaction follows a second order kinetics; i.e the rate is dependent on the concentration of ester as well as the concentration of the base. The rate of the reaction is represented by the following equation

$$Rate = k\ [CH_3COOC_2H_5]\ [NaOH] \qquad\qquad .....(2.45)$$

The rate constant for the reaction is determined from the second order rate equation

$$k = \frac{2.303}{t(a-b)}\ log\ \frac{b(a-x)}{a(b-x)} \qquad\qquad .....(2.46)$$

where, a and b are the initial concentrations of the alkali and ester in the reaction mixture.

a – x = Amount of sodium hydroxide present at time t in 20 ml reaction mixture

a – b = Excess of sodium hydroxide over ester

b – x = Amount of ester present at time t

Sodium hydroxide is consumed in the reaction as the reaction progresses. So, the rate of the reaction is monitored by withdrawing samples from the reaction mixture in a definite time intervals. The withdrawn samples are neutralized with standard hydrochloric acid and the excess acid is titrated back with sodium hydroxide using phenolphthalein indicator.

Procedure for kinetics study

Take ethyl acetate solution (100 ml, 0.01M in water) and sodium hydroxide (100 ml, 0.02M) solution in two different conical flasks and put the solutions in a thermostat at 30 ^{0}C. Mix the two solutions and immediately withdraw 20 ml of samples. Add hydrochloric acid (20 ml, 0.02M) to the sample. Titrate the excess acid in the solution with sodium hydroxide (0.02M) using phenolphthalein indicator. Repeat this procedure after every five minutes time intervals. Allow the reaction mixture to stand for 24 h to complete the reaction. Determine the final titre value by withdrawing 20 ml samples from this mixture as before. Thus, knowing a, b, a–x, b–x and a–b; k can be determined by graphical method by using equation 2.46.

Iodination of acetone

Many reactions follow zero order kinetics. Some examples of such reactions are heterogeneous reactions and enzyme reactions. However, there are some organic reactions that follow zero order kinetics under special conditions. One such example is, iodination of acetone in the presence of an acid. Iodine reacts with acetone in the presence of mineral acids to give iodo-acetone and hydrogen iodide as shown below

$$CH_3COCH_3 + I_2 \xrightarrow{H^+} CH_3COCH_2I + HI \qquad .....(2.47)$$

The rate of the reaction can be followed by recording the decrease in the concentration of iodine in the reaction mixture over a period of time.

$$\text{Rate} = - d[I_2]/dt \qquad\qquad\qquad \dots\dots(2.48)$$

Therefore, the rate law is:

$$- d[I_2]/dt = k\,[H^+]^y\,[(CH_3)_2C = O]^z \qquad\qquad \dots\dots(2.49)$$

The reaction will follow zero order kinetics in the presence of excess of acid as well as acetone. For a small change in concentration of these species the overall concentration will be large enough and apparently no changes will be observed. The kinetics of the reaction is monitored by withdrawing a known volume of the reaction mixture at a definite time interval and titrating the amount of liberated iodine. The iodine is estimated by conventional titration with standard sodium thiosulphate solution.

Procedure

Take acetone (10 ml), sulphuric acid (20 ml, 0.5M) and distilled water (60 ml) in a conical flask. Take a saturated iodine solution (50 ml) and the acetone mixture separately, and put them over a thermostat at 30 ^{0}C. After an equilibrium temperature is reached, add iodine solution (10 ml) to the acetone mixture. Withdraw 10 ml of the reaction mixture in every five minutes intervals of time and add sodium acetate (10 ml, 0.01M) to each of the sample to stop the reaction. Titrate the solutions with standard sodium thiosulphate solution (0.01M) using starch as indicator. Plot time (abscissa) vs. volume of sodium thiosulphate consumed at each time interval. A straight line will be obtained. Rate constant k can be calculated from the slope.

Study of redox couples

Metal ions can form their compounds at different oxidation states. Some of the oxidation states of metal ions are stable and some of them are not. For example, among oxidation state of copper, stable state are +2 and zero, the +1 oxidation state is not very stable. The copper compounds at +1 oxidation state disproportionate to give copper at zero and copper compounds at +2 oxidation   state.   There   are   many   other couples such as Fe^{2+} / $Fe^{3+,}$ Cr^{2+} / Cr^{3+}, Co^{2+} / Co^{3+}, Hg^+ / Hg^{2+} etc are associated with one electron difference but there are other couples like Sn^{2+} / Sn^{4+}, Pb^{2+} / Pb^{4+}, Mn^{2+} / Mn^{7+}, Cr^{3+} / Cr^{6+} which has differences of more than one electrons in the different oxidation state. Inter-conversions among these oxidation states are very important in redox reactions. Some of these redox processes are reversible and can be either caused by electrochemical means or by chemical means. Thus, when such metal ions with redox couples are treated with appropriate organic or inorganic substrates, they can lead to oxidation or reduction of the organic or inorganic substrates. Such redox couples may be associated with organic substrates, where radicals may be reversibly generated. Moreover, there are possibilities of step wise oxidation or reduction of organic or inorganic substrate. These facts can be studied by a simple technique, which is called cyclic voltametry. Cyclic voltametry helps in understanding a redox couple.

In this electrochemical technique the potential is swept from a resting potential (measured against a standard electrode, generally calomel electrode) at a constant rate to a limiting potential and returns to the original potential. Thus, any reversible redox process falling within the sweeping area will show its characteristic peak. The current at each increment and decrement in voltage is plotted. These types of plots give the anodic and cathodic peaks for each reversible couple, with a separation of the current peaks or depreciation at the oxidation and reduction voltages. If the Ip_a is anodic current and Ip_c is cathodic current for a redox process; then the ratio Ip_a/Ip_c will be close to unity for a reversible couple. Moreover, the difference between the anodic and cathodic e.m.f maximum will be less than 60 mV for a reversible process. The numbers of electron transferred during oxidation and reduction is however to be determined by a separate experiment called coulometry. The cyclic voltametry is generally performed in organic solvent or water with low concentration of substrates, so the conductivity of the solutions is low. In order to increase the efficiency of electron transfer process, a soluble quaternary ammonium salt is added as a supporting electrolyte to a solution. Generally tetrabutyl ammonium perchlorate is the choice as it does not undergo electrochemical transformation in a wide voltage range. The electrochemical properties are important in understanding redox properties of an electroactive compounds, it also allows us to know about feasibility of redox reactions. The study of redox processes of 1,4-benzoquinone by cyclic voltametry is interesting to understand its electrochemical properties. The 1,4-benzoquinone has two reversible one electron redox processes as illustrated in scheme 2.4:

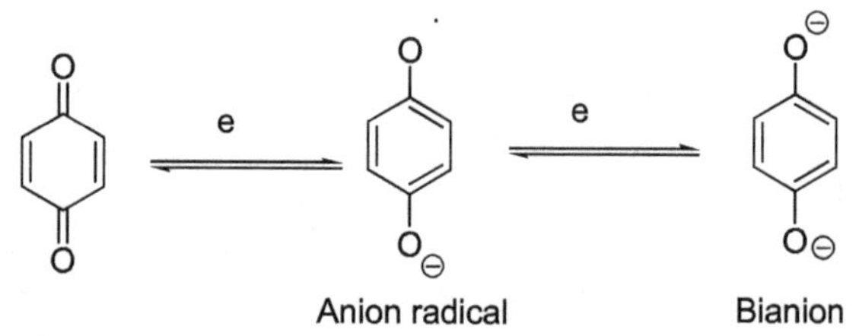

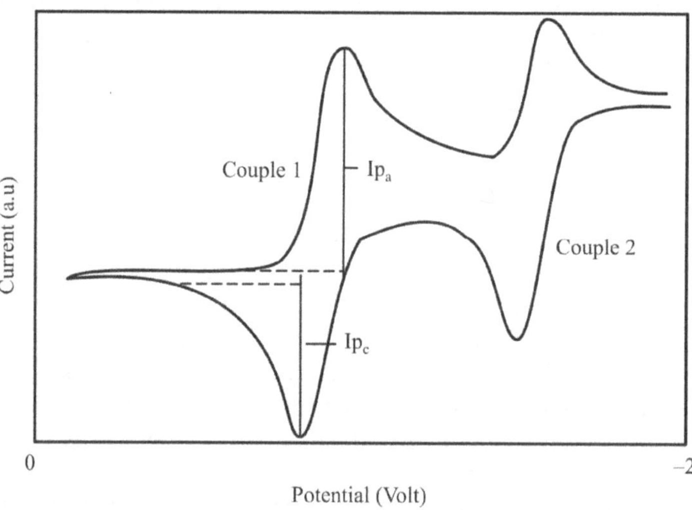

Scheme 2.4 The anion radical and bianion of benzoquinone under electrochemical condition

These redox couples can be studied by cyclic voltametry and both the reversible cycles appear at two distinct positions. The cyclic voltamogram for 1,4-benzoquinone in the negative side of the scale 0 to -2V in acetonitrile in the presence of tetrabutyl ammonium perchlorate with a sweep rate 100 mv/sec using an Ag/AgCl and platinum electrode with Calomel as standard electrode is shown in Fig. 2.3.

Fig. 2.3 The cyclic voltamogram of benzoquinone in acetonitrile with 100 mV per second scan speed using tetrabutylammonium perchlorate as supporting electrolyte.

The first couple corresponds to formation of an anion radical and the couple 2 corresponds to bianion formation. Similarly, the redox process of 1,4 diamino benzene can be studied by cyclic voltametry. In this case the electrochemical processes are as follows:

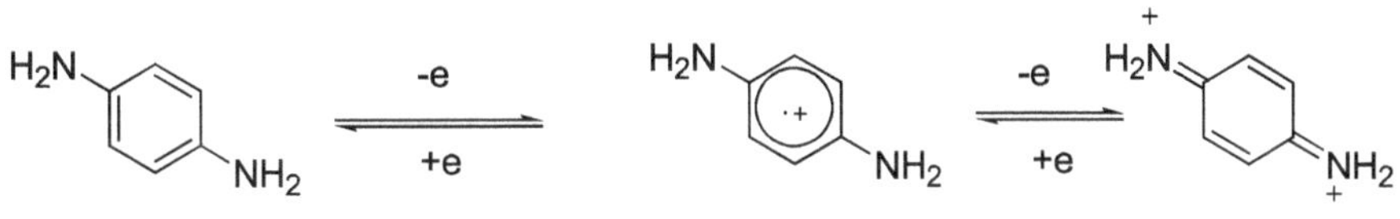

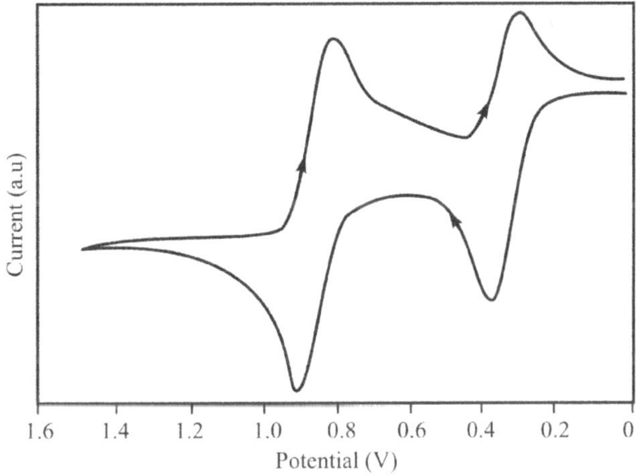

Scheme 2.5 Redox couples in the 1,4-diaminobenzene

These two couples appears in cyclic voltamogram as shown in Fig 2.4.

Fig. 2.4 The cyclic voltamogram of 1,4-diaminobenzene in acetonitrile with 100 mV per sec scan speed using tetrabutylammonium perchlorate as supporting electrolyte.

It is important to prepare various derivatives of 1,4-benzoquinone to understand electrochemical properties and also to determine their applicability in photo-oxidation reactions. The amino and thiolato derivatives of benzoquinones can be easily prepared. The 2,5 *bis*-(alkyl or arylamino) 1,4-benzoquinones are such derivatives and are synthesized by reactions of various amines with 1,4-benzoquinone in alcohol. The reaction does not require additional reagent. For example, 2-propylamine reacts with methanolic solution of 1,4-benzoquinone to give 2,5-*bis*(2-propylamino)-1,4-benzoquinone (equation 2.50).

$$2\,RNH_2 \; + \quad \text{(1,4-benzoquinone)} \quad \longrightarrow \quad \text{(product)} \qquad R= \text{isopropyl} \quad .....(2.50)$$

Preparation of 2,5-*bis*(2-propylamino)-1,4-benzoquinone

Add 2-propylamine (0.6 gm, 10.2 mmol) dropwise to a well-stirred solution of 1,4-benzoquinone (0.54 gm, 5 mmol) in methanol (15 ml) at room temperature. The yellow color due to the parent quinone disappears slowly and the solution turns black. Stir the solution for 15 hrs at room temperature. Make the volume of the solution to half under reduced pressure. A brown precipitate will appear. Filter the precipitate and dry it at room temperature and further recrystallize it from methanol to obtain 2,5-*bis*(2-propylamino)-1,4-benzoquinone. Melting point: 185^0C. ^{1}H NMR (CDCl$_3$, 400 MHz, ppm): 1.32 (d, J = 6.8 Hz, 12H), 3.65 (heptet, J = 6.8 Hz, 2H), 6.50 (broad s, 2H), 6.70 (s, 2H). IR(KBr, cm^{-1}): 3210 (s), 2976 (s), 1564 (s), 1487 (s), 1314(s). UV–vis (χ_{max}, CH$_3$CN) 341 nm (ε = 15,387.6 cm^{-1} M^{-1}).

The compound shows a reversible couple for generation of anion radical:

$$\mathbf{A} \quad \underset{}{\overset{e^-}{\rightleftharpoons}} \quad \mathbf{B} \qquad .....(2.51)$$

E$_{1/2}$(V)	i$_p$/i$_c$	ΔEp(V)
−1.0415	1.10	0.167

Partition coefficient

Separation of a particular compound from a mixture of compounds by solvent extraction is very important in natural product chemistry, as well as in synthetic chemistry. The knowledge on partition coefficient helps to understand many equilibrium processes that are useful in separation of components from a mixture of compounds. When a solute is allowed to dissolve in two non-miscible solvents, the solute distributes itself in two solvents in such a way that the ratio of concentrations of the solute in these two solvents is constant at a constant temperature. This happens when the solute dissolves in both the solvents remain in the same form. No chemical reaction should take place between solute and solvent, and there also should be no association or dissociation of the solute in these solvents. The partition coefficient is mathematically expressed as

$$K_d = \frac{C_1}{C_2} \qquad \qquad \ldots(2.52)$$

where C_1 and C_2 are the concentrations of the solute in the two different solvents. K_d is called the distribution or partition co-efficient.

Determination of partition coefficient

Let us look at the distribution of iodine between water and benzene. Experiment can be done as follows:

Take five independent solutions in five 250 ml stoppered bottles as following :

25 ml of saturated iodine solution (in benzene) and 5 ml of pure benzene

20 ml of saturated iodine solution (in benzene) and 10 ml of pure benzene

15 ml of saturated iodine solution (in benzene) and 15 ml of pure benzene

10 ml of saturated iodine solution (in benzene) and 20 ml of pure benzene

5 ml of saturated iodine solution (in benzene) and 25 ml of pure benzene

Add distilled water (50 ml) to each of these bottles. Shake the bottles for 2 hrs on a shaker or by manual stirring. Allow the bottles to stand, so that the liquids in the bottles separate into two distinct layers. Find out the concentration of iodine in both the layers of each bottle by taking out known volume of solutions and titrate them against the standard sodium thiosulphate solution (0.01M). Since the solubility of iodine in the two solvents is different; the volume of solution sodium thiosulphate will have to be adjusted for each titration. At the end of the experiment, the concentrations are to be made uniform for calculations. Find out the ratios of the concentrations of iodine in water and benzene in each case. It will be observed that in all the cases the ratios remains constant, within the limit of experimental error.

Study on the formation of potassium triiodide

The potassium iodide with iodine remains as potassium triiodide when dissolved in water, and such equilibrium can be studied by partition co-efficient method. Determining the partition coefficient of iodine between carbon tetrachloride and water in the presence and absence of potassium iodide helps in studying such equilibrium process. When iodine is added to a solution of potassium iodide, the complex potassium triiodide is formed and the following equilibrium is established:

$$KI + I_2 \rightleftharpoons KI_3 \qquad \qquad .....(2.53)$$

The equilibrium constant K of this reaction is given by

$$K = \frac{[KI_3]}{[KI][I_2]} \qquad \qquad .....(2.54)$$

where $[KI_3]$, $[KI]$ and $[I_2]$ represents the concentration of potassium triiodide, potassium iodide and iodine after the equilibrium is reached. At a constant temperature partition coefficient (K_d) is constant and the equilibrium constant of the reaction can be calculated.

After equilibrium is reached the concentration of free iodine cannot be determined by direct titration with sodium thiosulphate. This is because if iodine is removed by titration with sodium thiosulphate the equilibrium between iodine and potassium iodide will shift and more potassium triiodide will break into potassium iodide and iodine. To find out the equilibrium constant the concentration of iodine that has formed potassium triiodide will have to be found out by finding out the partition coefficient of iodine alone in a mixture of water and carbon tetrachloride. The potassium triiodide and potassium iodide are ionic compounds; they completely dissociate in the aqueous layer. These ionic compounds donot enter into the carbon tetra chloride layer. So an iodine solution is prepared in aqueous solution of potassium iodide. This solution is added to carbon tetrachloride while doing this experiment. The amount of iodine in water and carbon tetrachloride layer are determined by titrating a known volume of solution with standard sodium thiosulphate solution.

$$K_d = \frac{\text{Conc of free iodine in } CCl_4 \text{ layer}}{\text{Conc of free iodine in water layer}} \qquad .....(2.55)$$

Concentration of free I_2 in aqueouslayer

$$= \frac{\text{Conc of free iodine in } CCl_4 \text{ layer}}{K_d} \text{ g moles / lit}$$

$$= C_1/K_d \qquad \qquad .....(2.56)$$

Let the total concentration of iodine i.e. free $I_2 + KI_3$ in aqueous layer is C_2 g moles/lit

At equilibrium the concentration of the potassium triiodide in the aqueous layer

$$= (C_2 - \frac{C_1}{K_d})$$

The potassium triiodide is formed at the expense of an equivalent amount of potassium iodide from the aqueous layer. Hence, the concentration of potassium triiodide is directly proportional to the decrease in the concentration of potassium iodide in the water layer at the time of equilibrium.

Therefore, the decrease in the concentration of potassium iodide in aqueous layer is

$$= (C_2 - \frac{C_1}{K_d}) \quad \text{g.mole/lit} \qquad \qquad .....(2.57)$$

After the equilibrium is reached concentration of

$$[KI] = [C_{KI} - (C_2 - \frac{C_1}{K_d})] \qquad \qquad ......(2.58)$$

where C_{KI} is the initial concentration of potassium iodide

$$K = \frac{[KI_3]}{[KI][I_2]} = \frac{(C_2 - \frac{C_1}{K_d})}{[C_{KI} - (C_2 - \frac{C_1}{K_d})]\frac{C_1}{K_d}} \qquad \qquad .....(2.59)$$

Experimental procedure

Prepare the following solutions in five 250 ml bottles:

1. Potassium iodide solution (25 ml, N/10)
2. Potassium iodide solution (20 ml, N/10) and 5 ml water
3. Potassium iodide solution (15 ml, N/10) and 10 ml water
4. Potassium iodide solution (10 ml, N/10) and 15 ml water
5. Potassium iodide solution (5 ml, N/10) and 20 ml water

Place saturated iodine solution in carbon tetrachloride (30 ml) in each bottle. Shake the bottles by closing the mouths for 20 minutes. Determine the concentrations of iodine in all the layers.

Find out the partition coefficient K_d for iodine in water and carbon tetrachloride mixture by making four sets of solutions

1. Water (25 ml) + saturated I_2 in carbon tetrachloride (5 ml)
2. Water (20 ml) + saturated I_2 in carbon tetrachloride (10 ml)
3. Water (15 ml) + saturated I_2 in carbon tetrachloride (15 ml)
4. Water (10 ml) + saturated I_2 in carbon tetrachloride (20 ml)

Shake these solutions properly and determine the ratios of concentration of iodine by titrating with sodium thiosulphate in each layers.

Calculate the equilibrium constant by using equation 2.59.

Dimerisation of benzoic acid in benzene

Benzoic acid forms hydrogen bonded dimer in benzene. Whereas, benzoic acid is inter-molecularly hydrogen bonded in water. This phenomenon can be determined by studying distribution of benzoic acid between water and benzene. The concentration of benzoic acid in benzene and water layers can be determined by titrating with sodium hydroxide using phenolphthalein as indicator. When benzoic acid is allowed to distribute between benzene and water major portion of the benzoic acid dissolves in benzene. Two types of equilibrium exist in benzene and water layers. One is between the monomers of benzoic acid in benzene and water layer; and other is between the monomers and the dimers in the benzene layer.

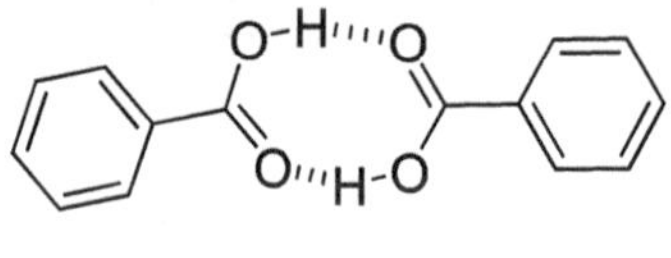

Fig. 2.5 Hydrogen bonded dimer of benzoic acid.

The equilibriums are

$$A \leftrightarrow 2A \leftrightarrow A_2$$
$$C_1 \quad C_A \quad C_A^2 \qquad \qquad .....(2.60)$$

Let C_1 and C_A be the molar concentration of the monomer in water and benzene layers respectively; and C_A^2 be the molar concentration of the dimer in benzene layer.

In this case the ratio of the total concentration of the benzoic acid in the two layers is not constant. The ratio of the concentration of the single benzoic acid molecules in the two liquid layers are constant, and this ratio is the partition co-efficient of the benzoic acid in water and benzene.

$$K_d = C_A/C_1 \qquad \qquad \dots\dots(2.61)$$

Consider the equilibrium between the monomers and dimers in the benzene layer,

$$2A \leftrightarrow A_2 \qquad \qquad \dots\dots(2.62)$$

$$C_A \qquad C_A^2$$

Equilibrium constant

$$K_c = C_A^2/(C_A)^2$$

$$C_A = \sqrt{(C_A^2)}/\sqrt{K_c} \qquad \qquad \dots\dots(2.63)$$

Combining the equations 2.61 and 2.62

$$\frac{C_A}{C_1} = \frac{\sqrt{C_A^2}}{\sqrt{K_C \times C_1}} = K_d$$

$$\frac{\sqrt{C_A^2}}{C_1} = K_d \times \sqrt{K_C} = K'(\text{constant}) \qquad \dots\dots(2.64)$$

Since most of the benzoic acid remains in the dimeric state in the benzene layer, C_A^2 may be taken as the total molar concentration of dimeric form of benzoic acid in the benzene layer.

If C_B and C_w be the total concentration of benzoic acid in benzene and water respectively

$$C_A^2 = C_B/2 \quad \text{and} \quad C_1 = C_w$$

From equation 2.64

$$\frac{\sqrt{C_B}}{\sqrt{2}\times C_W} = K' \qquad\qquad .....(2.65)$$

$$\frac{\sqrt{C_B}}{C_W} = K'\times\sqrt{2} = K'(\text{cons}\tan) \qquad\qquad .....(2.66)$$

The experiment can be done by distributing benzoic acid in different compositions of water and benzene and estimating the concentration of benzoic acid in each layer.

Hydrolysis equilibrium

Partition coefficient can be used to study hydrolysis of an ammonium hydrochloride. Consider the reaction of anilinium hydrochloride undergoing hydrolysis to aniline.

$$.....(2.67)$$

The hydrolysis equilibrium is

$$K_h = \frac{[\text{Anilline}][\text{HCl}]}{[\text{Aniline hydrochloride}]} \qquad\qquad .....(2.68)$$

To study this reaction the concentration of aniline with respect to anilinium hydrochloride is to be determined. Aniline is a neutral compound; whereas its hydrochloride is ionic. Thus, they can be distributed between benzene and water, so that partition coefficient is

$$Kp = \frac{\text{Concentration of aniline in benzene}}{\text{Concentration of anilinium hydrochloride in water}} = \frac{C_{anil}}{C_{anilhydr}}$$

$$.....(2.69)$$

Estimation of aniline in benzene layer will give the concentration of hydrolysed aniline and this will be equivalent to the hydrochloric acid in water layer. Aniline can be estimated by bromination method; by using potassium bromate, potassium bromide, and hydrochloric acid. Brominating mixture allows generation of bromine in situ, which is added to aniline to give tribromo benzene anilinium bromide.

$$KBrO_3 + KBr + HCl \rightarrow Br_2 + H_2O + KCl \qquad .....(2.70)$$

$$.....(2.71)$$

The excess bromine in the solution is estimated by a back titration by treating the solution with potassium iodide. Which liberates iodine by oxidizing potassium iodide which is added in known quantities from external source, and the liberated iodine is estimated by sodium thiosulphate solution. Carrying out the experiment with a constant concentration of aniline hydrochloride and titrating the aniline in benzene layer with 1hr time interval will give the amount of aniline hydrochloride hydrolysed and from these titrations the plot of concentration of aniline formed with time is plotted.

Molecular weight of polymers

Molecular weight determination is an important issue in chemistry. For polymers, however molecular weights are spread over a range. Thus, in such cases the average molecular weights are determined. The molecular weight of a polymer can be determined by various techniques such as viscosity, refractive index etc.

Molecular weight of a soluble polymer can be easily determined by measuring the viscosity of the polymer in solutions. The viscosity of solution of a polymer is given by Mark-Kuha Houwink equation that relates the intrinsic viscosity and molecular weight of a polymer as

$$[\eta] = K\, M^a \qquad .....(2.72)$$

where, $[\eta]$ is the intrinsic viscosity. M is the average molecular weight of the polymer. K and a are two constants for a particular solvent polymer system at a constant temperature. For solvents, a value of a is indicative of the suitability of the solvent for molecular weight determination. A value of $a = 0.8$ is typical for good solvents. For most of the flexible polymers, a is $0.5 \leq a \leq 0.8$. For semi-flexible polymers, $a \geq 0.8$. For polymers with an absolute rigid rod like structure, $a = 2.0$.

Intrinsic viscosity is a measure of contribution of a solute to the viscosity η of a solution. It is defined as

$$[\eta] = \lim_{c \to 0} \frac{\eta_{sp}}{C} = \lim_{c \to 0} \frac{\eta_{solute} - \eta_{solvent}}{C} \qquad .....(2.73)$$

where, C is the concentration of polymer in gm per 100 ml of solvent.

Intrinsic viscosity is generally obtained by plotting η_{sp}/C vs concentration (abscissa). The extrapolation of η_{sp}/C to $C = 0$ gives the intercept of the curve. From this intercept the intrinsic viscosity can be determined.

Determination of intrinsic viscosity

Experimentally intrinsic viscosity is determined by using Ostwald's viscometer. The molecular weight of polystyrene is determined by viscocity measurement as follows

Prepare a solution of polystyrene (0.5 gm) in toluene (25 gm) and take 10 ml of the solution in a viscometer thermostated at 25^0C. Determine the time of flow of the solution through the viscometer. Withdraw the solution from the viscometer and take equal volume of the solvent and determine the time of flow. Prepare two or three solution at different concentrations of polymer solution in toluene and determine the time of flow of 10 ml each through the viscometer. Determine densities of all the solutions along with the solvent by using density bottle.

Viscosity from the experiment is calculated by using the following equation

$$\frac{\eta}{\eta_0} = \frac{\rho t}{\rho_0 t_0} \qquad .....(2.74)$$

where, t and t_0 are time of flow of the solution and the solvent and ρ and ρ_0 are the respective densities.

The values of the K and a, can be determined for a sample of known molecular weight by using the equation

$$\log [\eta] = \log K + a \log M \qquad \qquad .....(2.75)$$

A plot of log [η] vs log M gives a straight line with slope equal to a and intercept is equal to log K.

Gel permeable chromatography

A fairly broad distribution in molecular weight is common feature of polymers. This type of distribution affects the viscosity and other properties such as miscibility, and modulus of elasticity of a polymer. By understanding the distribution of molecular weight profile versus number of fractions provides insight to the molecular weight of the polymers. The Fig. 2.6 shows a biomodal i.e. two identifiable peaks in a GPC profile. These patterns are representation of skewed distribution. The skewed distribution means that the distributions are not symmetric about their mean. The unimodal (one peak) symmetric distribution is easy to describe and involves two parameters the mean, μ and the standard deviation, σ.

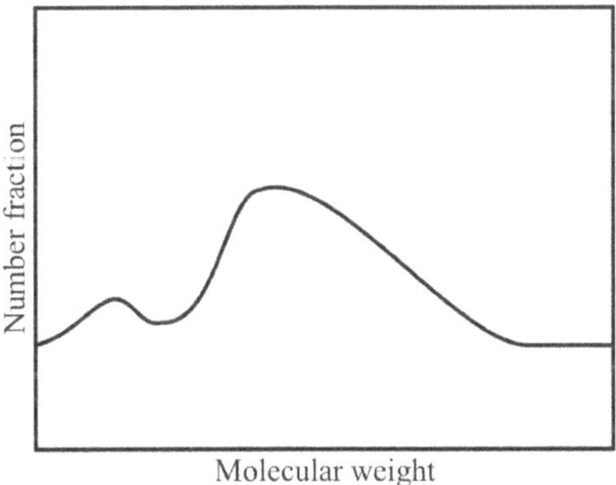

Fig. 2.6 A distribution of molecular weight in a polymer

The number average molecular weight, M_n is

$$M_n = \frac{\Sigma N_i M_i}{\Sigma N_i} \qquad \qquad .....(2.76)$$

where, N_i is the number of molecules having molecular weight M_i. The weight average molecular weight (mass average), M_w, is given by:

$$M_w = \frac{\Sigma N_i M_i^2}{\Sigma N_i} \qquad \qquad .....(2.77)$$

The polydispersity index, M_w/M_n, is the ratio of the second moment to the square of the first moment about the origin. For a simple unimodal symmetric distribution only these definitions are good enough to describe the system, however, for skewed distributions more complicated parameters are to be added.

For determination of molecular weight two types of techniques are used. These are called primary and secondary techniques based on whether the measurement is based on calibrated samples or not. The primary techniques are based on colligative properties of the polymer. The melting point depression, vapor pressure and osmotic pressure are used for such measurements. Light scattering and chemical analysis of end groups and viscocity measurements are used for determination of molecular weight of polymers. The other secondary techniques are gel permeation chromatography (GPC) technique. GPC results in measurement of the entire molecular weight distribution for any type of molecular weight distributions. For GPC measurement, it is usually calibrated using mono-disperse standards, most commonly polystyrene standards. The molecular weight is then expressed as equivalent to polystyrene molecular weight.

In gel permeation chromatography a dilute polymer solution is passed through a tubular column packed with polymeric gel beads. Some of the polymer chains are forced into the pores of the gel under such high-pressure flow, while the others pass by the gel beads. The retention time of a polymer in the packed column thus depends on the path through the gel. For example, a polymer with small molecular weight will pass easily and will experience the routine adsorptions. Whereas, a high molecular weight polymer will not match the pore size, thus they will be excluded, and pass at a higher speed directly to the exit portion of the column. So, the detector will be able to recognize each of them. Refractometers are generally used as detectors. The output of a GPC reflects the number of chains of a particular set of molecular weight distribution.

Dilute solution of polymer is injected into the gel permeation chromatograph. The pump sends the solvent through the column, which also carries the injected sample. Once the sample reaches the detector it notes the overall concentration of polymer in the eluted solvent as a function of time at constant volumetric flow rate. The elution volume is the amount of solvent flown at a constant flow rate for a constant time through the column. The time that the polymer takes to elute from the column is called the retention time, t_R, and the elution volume for this time is called the retention volume, designated as V_R. With the use of an appropriate gel in the column for a molecular weight range under consideration, the relation between V_R and molecular weight is given by $V_R = V_{R,0} - kM$, where M is the molecular weight, k and $V_{R,0}$ are constants. $V_{r,0}$ is constant for a particular polymer/solvent/gel system. These constants are determined by eluting two or more mono-disperse standards.

Molecular weight from freezing point depression

The freezing point depressions can be used to determine molecular weight of a compound. When certain amount of a non-volatile, non-electrolyte is dissolved in a solvent, the freezing point of the solvent is lowered. For a dilute solution, the depression of the freezing point is given by Raoult's law.

$$\Delta T_f = K_f\, m \qquad\qquad .....(2.78)$$

where, ΔT_f = Depression of freezing point = Freezing point of the solvent – freezing point of the solution

and K_f = Constant for a solvent, m is the molality of the solution

$$K_f = \frac{RT_o^2}{1000\,L_f} \qquad\qquad .....(2.79)$$

T_0 = Freezing point of the pure solvent

L_f = Latent heat of fusion per gram of the solid solvent

$$= \frac{W_2 \times 1000}{M_2 \times W_1} \qquad\qquad .....(2.80)$$

W_2 = Mass of the solute dissolved.

W_1 = Mass of the solvent taken for making the solution.

Hence equation 2.79 may be written as

$$\Delta T_f = \frac{RT_o^2}{1000\ L_f} \times \frac{W_2 \times 1000}{M_2 \times W_1}$$

$$M_2 = \frac{RT_o^2\ W_2}{L_f W_1 \Delta T_f} \qquad\qquad .....(2.81)$$

Measuring the freezing point of the solvent and the solution independently the depression of freezing point can be determined. Once the ΔT_f is experimentally determined then the molecular weight M_2 of the solvent can be found out from the equation 2.81.

Procedure

Determine the freezing point of a weighed amount of solvent. Take an exactly weighed amount of solid whose molecular weight is to be determined; and dissolve the solid in the pure solvent where the compound is soluble and whose freezing point depression constant is known. Determine the freezing point of the solution. Find out the difference of the two freezing points. Calculate the value of M_2 from equation 2.81. The latent heat of fusion for the solid is determined by an independent experiment based on calorimetric method.

Alternatively, the K_f of the solvent is determined by measuring the freezing point depression of a sample whose molecular weight is known.

Adsorption

When a gas or liquid or solute accumulates on the surface of a solid, or a liquid as a film of molecules or atoms it is called adsorption. It occurs in many natural physical, biological, and chemical systems. It is widely associated with activated charcoal, resins, and porous materials. The absorption property is routinely used in water purification. It has important role in heterogenous catalysis. In chromatographic techniques the role of adsorption is indispensable. Adsorption occurs due to surface energy of absorbents. Atoms positioned on a surface of an adsorbent are not wholly surrounded by other atoms and therefore due to non uniform forces it is attracted to the adsorbents. The nature of the bonding of the absorbents depends on the property of the entity under consideration. Depending on the strength of binding the adsorption process is generally classified as physisorption and chemisorption. Physisorption is characterized by weak Van-der Waals forces whereas the chemisorption has characteristics that is close to covalent bonding.

Let us take activated charcoal as an example to illustrate adsorption. Charcoal is a highly porous, amorphous solid consisting of micro-crystallites. These microcrystalline solids have graphite type lattices. That is to say that it has sp^2-hydridised carbons; connected to each other forming layered structures. Charcoal is usually available in the form of small pellets or powder. Activated charcoal is manufactured from coal, peat, wood, nutshells etc. The charcoal manufacturing is done by carbonization process by drying and then heating. The heating is done at 400–600°C in an oxygen-deficient atmosphere to separate by-products such as tars and hydrocarbons. The charcoal, thus obtained is activated by exposing to an oxidizing agent. Generally, steam or carbon dioxide at high temperature is used for the activation. The activation process leads to porous, three-dimensional lattice structure of graphite in the charcoal. The size of the pores depends on the conditions applied during the activation process. For example longer exposure times result in larger pore sizes.

Adsorption isotherm of acetic acid on charcoal

Adsorption is governed by different equations, which are referred to as isotherms. These isotherms are formulated on the basis of the ways in which the layers of adsorbed species are placed over adsorbent and their strength of interactions at a constant temperature. For adsorption of a dissolved substance by solid surfaces, H. Freundlich gave the following relationship as

$$x/m = KC^{1/n} \qquad \qquad \text{..... (2.82)}$$

where x is the amount of the substance adsorbed, m is the mass of the adsorbent, and C is the equilibrium concentration of the dissolved substance, K and n are empirical constants. These constants depend on the nature of the absorbent, the adsorbed species and also on temperature.

Taking logarithm of both sides of equation 2.82, we have

$$\text{Log } (x/m) = \log K + 1/n \log C \qquad \qquad \text{.....(2.83)}$$

Hence a plot of log x/m vs log C should be a straight line with slope equal to 1/n and the intercept equal to log K.

Experimental Procedure

Prepare a standard solution of oxalic acid (250 ml, 0.1N). Prepare also a solution of sodium hydroxide (500ml, 0.1N) and standardize it accurately with standard oxalic acid using phenolphthalein indicator. Prepare another solution of acetic acid (500 ml, 0.1N) and standardize it also with standard sodium hydroxide using phenolphthalein as indicator. Label six clean and dry bottles (200 ml capacity) as 1, 2, 3, 4, 5, 6. Place accurately weighed (~2 g) powdered charcoal in each bottle. Now place 100 ml, 75 ml, 50 ml, 40 ml, 20 ml and 10 ml acetic acid solution in the bottles labeled as 1, 2, 3, 4, 5, 6 respectively. Make up the volume of each bottle to 100 ml by adding 0, 25, 50, 60, 80 and 90 ml of water in bottles labeled 1, 2, 3, 4, 5, 6 respectively. Stopper the bottles tightly. Shake the bottles. Filter the solutions separately with dry filter paper placed fitted in dry funnels. Reject the first 15 ml solution from each solution. This is done to eliminate error due to adsorption of acid by filter papers. Collect the filtrates separately in dry beakers. Determine the concentration of acetic acid in each bottle by titrating a known volume of solution against standard sodium hydroxide solution using phenolphthalein indicator. Note the temperature before and after the experiment.

Plot log x/m as ordinate against log C as abscissa and find the value of K and n from the graph.

Alloy

Alloys are prepared commercially by melting active metals. Various alloys are prepared for different purposes. For example brass is used for utensils, steel for higher strength materials, Raney nickel for hydrogenation catalyst, lead-tin alloy for solder etc. The phase diagrams are useful in understanding compositions and conditions for co-existence of different phases of metals of alloys. As an illustrative example, consider the alloy of lead and tin, cooling curves for pure lead and tin, as well as alloy of lead and tin with 32:68 ratio are shown in Fig. 2.7. One can see that each of the three materials have characteristic melting points.

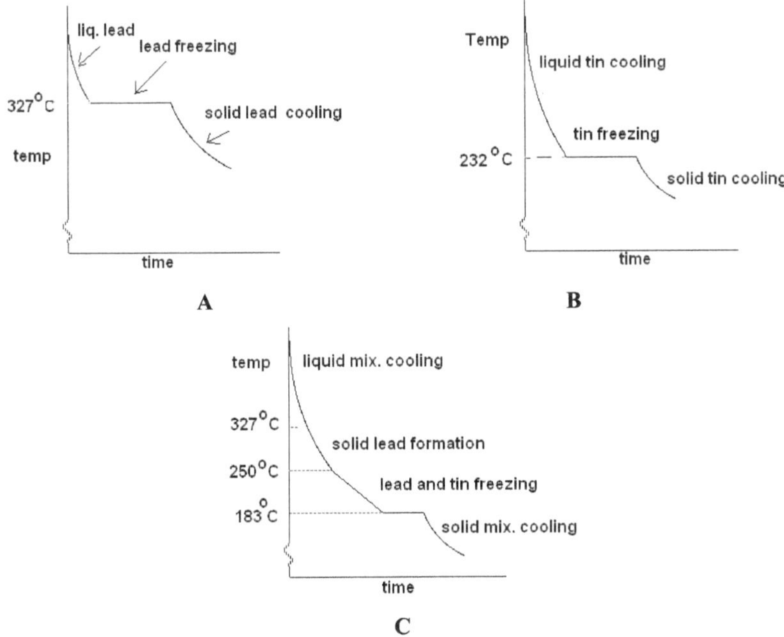

Fig. 2.7 Cooling curve for A) Lead, B) Tin, C) Lead-tin alloy

Preparation of alloy of lead and tin

Place lead (0.32 mole) and tin metal (0.68 mole) in a silica crucible. Melt the metals in the silica crucible over a flame. Cool the melt and remove the alloy in the solid form from the crucible.

Zeolites

Zeolites are three-dimensional alumino-silicates, built from AlO_4^- and SiO_4 tetrahedra. Zeolites are usually synthesized under hydrothermal conditions, from solutions of sodium aluminate, sodium silicate, or sodium hydroxide. The general formula of zeolite is $M_{x/n}[(AlO_2)x(SiO_2)y]$ with M as the compensating cation. It can be a quaternary ammonium ion with valence n. There are many forms of zeolites; each of them is designated by specific name. Among them ZSM-22 is a high-silica zeolite, it has unidimensional channels and medium-size pores of c.a. 0.45-0.55 nm constituting 10-membered aluminosilicate rings, provide special shape-selective properties.

Each of the zeolites have characteristic powder X-ray diffraction patterns. The powder X-ray pattern of ZSM-22 is shown in Fig. 2.8.

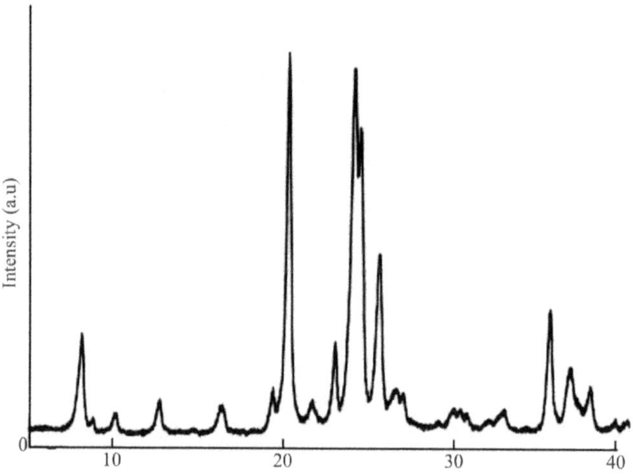

Fig. 2.8 Powder X-ray diffraction pattern of zeolite ZMS-22

Synthesis of zeolite is highly dependent on the use of different reaction parameters, as well as on the organic ammonium salts used in their synthesis.

Synthesis of zeolite ZSM-5

Take finely ground sodium hydroxide (0.510 g), silicic acid (2.01 g) and tetrapropylammonium bromide (1.01 g) in a beaker (250 ml). Mix them with water (5.0 ml) and n-propylamine (1.0 ml) and mix them together. Take a solution of aluminum sulfate (1.0 ml, 1M) with 0.05 ml of the concentrated sulfuric acid in another beaker (50 ml). Mix the two solutions. Add distilled water to make the volume to 25 ml, and stir for ten minutes. Put the solution in a pressure vessel and seal it. Heat the sample at 160 °C and for 48 hrs in the sealed tube. Cool the reaction mixture and filter the reaction mixture by using Buchner funnel. Wash it three times with small amounts of water and then dry it for 20 mins on a filter paper. Calcine the crude zeolite thus obtained, with increments of 50 °C to 100 °C. When water vapor is released, then heat the mixture with increments of 100 °C to 500 °C to remove the organic materials. Continue heating the mixture at 500 °C for 2 hrs. Sodium ions remaining in the zeolite can be ion exchanged for protons to fully convert the zeolite to its acid form. To do this, place the calcined product in a beaker (100 ml), and stir it with aqueous ammonium sulphate (12.60 ml of 1M) solution for 15 minutes. Filter the solid and wash it with acetone. Further to this wash it with water till zeolite is free from sulphate. This may be tested with barium(II) chloride solution, as presence of sulphate will lead to white precipitate. Dry the residue in an oven at 120°C for 30 minutes. Heat the ammonium zeolite for 3 hrs, so that all ammonium ions are removed. The zeolite will be obtained as off-white colored powder.

Indicators

Indicators are used in different volumetric estimation experiments and also to detect the pH of a solution. Depending on their nature of performances, they are differentiated in different categories. For example, acid-base indicator changes color in acid and base. A redox indicator changes color on electron transfer and accordingly, the oxidised or reduced form may have different colors. Thus, an electron transfer process can be detected by color changes. Whereas, adsorption indicators act on the principle of competition of adsorption processes between two or more adsorbents.

A common acid base indicator in neutral form may or may not have color but a color change takes place upon protonation or deprotonation. An indicator HI_n, on deprotonation gives I_n^- which may have different color from the parent compound. The difference in color generally observed due to changes of resonance structures. For example phenolphthalein on deprotonation adopts a highly conjugated structure and it shows pink color. The methyl orange can have quinonic structure on deprotonation which causes the color in basic medium.

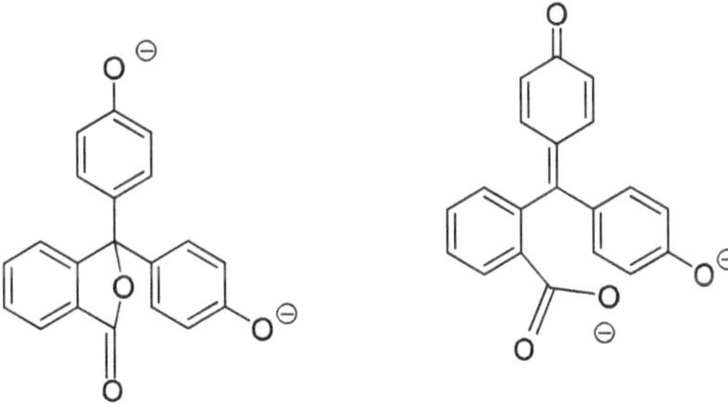

Phenolphthalein colorless   Phenolphthalein purple

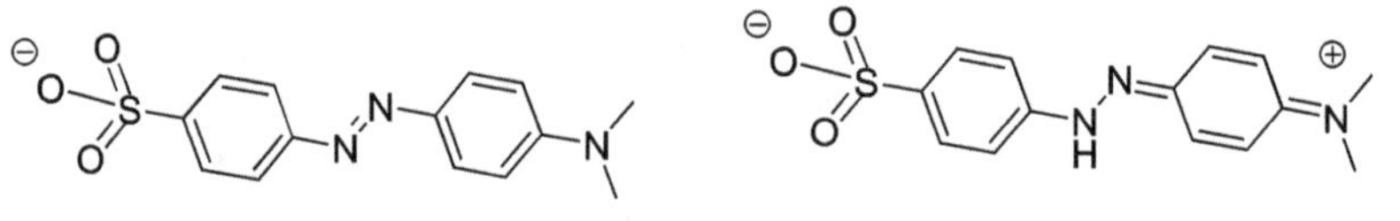

Yellow form of methyl orange

Red form of methyl orange

Fig. 2.9 Structures of some indicators

Flower coloration

Anthocyanins are colored compounds found in many plants or plant parts. The anthocyanins are red in acidic medium and blue in basic medium. Anthocyanins extracted from red cabbage leaves are used as crude acid-base indicator. Bright red, blue, or purple colored flower contains cyanidin pigment. The cyanidin pigment can exist in different forms depending on pH of a medium as illustrated below:

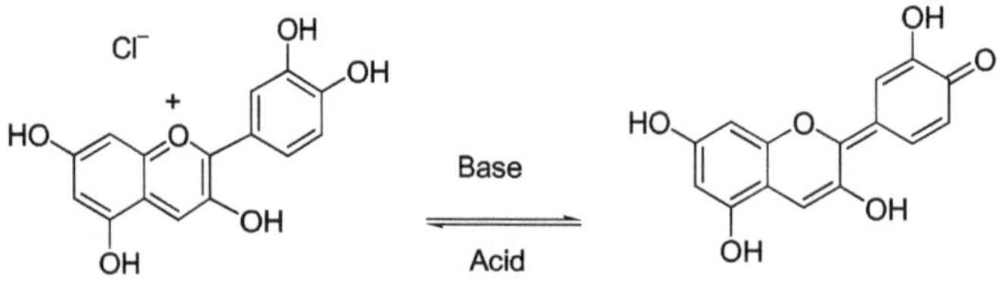

Base

Acid

Cyanidin chloride in acid

Cyanidin on deprotonation

$$.....(2.84)$$

The color of cyanidin in acidic medium is bright red. In basic medium the catechol frame gets deprotonated and transforms to a keto form. The keto form of cyanidin is blue or violet. A cyanidin molecule attached to sugar molecules is called anthocyan or anthocyanin. For example, the anthocyan that makes roses red has the following structure:

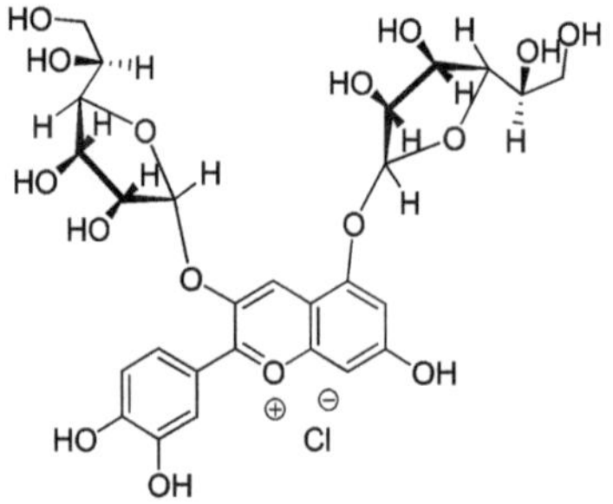

Fig. 2.10 Structure of anthocyanin

The role of the sugar in such biological system is to increase solubility of these compounds in water as well as to make specific supramolecular assemblies. The colors demonstrated by flowers are from similar pigments having common structural features. The fluids in the pigment-bearing tissues of different flowers have different pH values and these pH differences control the color of the flowers. Further to this coordination effect of such pigments to metal ions such as magnesium, calcium are resposible for certain colors.

Redox indicators

Diphenylamine is an example of a redox indicator. Diphenylamine gets oxidised to form conjugated structure as shown in scheme 2.6; the oxidized form shows the observed color change in a redox titration.

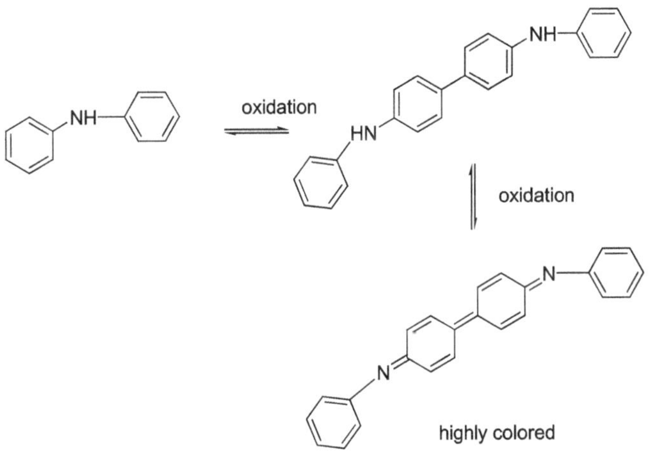

Scheme 2.6 Redox behaviour of diphenylamine

Complexometric titration

For complexometric titration the advantages of competitive binding abilities of a ligand and that of the indicator is taken. In these cases the complexes formed by the indicator with the metal ions have different color from a free metal ion or the color of the indicator. The criterions for such absorption phenomenon to show indicator property is that there should be difference in binding ability of two metals to the indicator and such binding should be reversible. Eriochrome black-T is such an indicator, which is used for estimation of calcium.

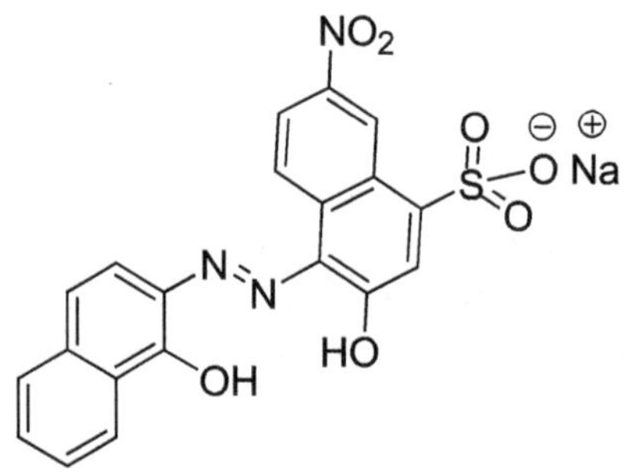

Fig. 2.11 Structure of eriochrome black-T

For understanding of complexometric titration consider the example of estimation of calcium in hard water. The hardness of water is generally due to dissolved calcium and magnesium salts mainly in the form of carbonates. The amount of calcium and magnesium ions can be determined by complexometric titration. In complexometric titration concentration of a metal ion is determined via complex formation by a strong ligand. In the estimation of magnesium and calcium, disodium salt of ethylenediamine tetraacetic acid is used. The equilibrium constant of these complexes plays a crucial role in these estimations. A metal ion in the form of a complex with an organic dye (indicator) is made initially which has a characteristic color. During the course of titration the ligand forms complex with the metal ion thereby setting the dye free from metal ion. At the neutralization point, when all the dye molecules get free, the solution assumes the color characteristic of the dye.

When calcium ions are titrated with di-sodium salt of ethylenediamine tetra-acetic acid (EDTA) a relatively stable complex is formed in basic medium as follows:

$$Ca^{2+} + H_2Y^{2-} = CaY^{2-} + 2H^+$$

where, $\quad\quad\quad\quad H_4Y = EDTA$ $\quad\quad\quad\quad\quad\quad\quad\quad\quad$ .....(2.85)

In alkaline medium the reaction is driven toward the forward direction. Eriochrome black T is used as indicator in this process. It is red when it forms complex with Ca^{2+}. At the end point a blue colored solution is observed.

Procedure

Take standard EDTA solution (strength to be decided by the amount of calcium in water sample) in a burette. Pipette out 25 ml of water sample into a 250 ml conical flask. Add 5 ml buffer solution (pH = 10) followed by 4 to 5 drops of Eriochrome black-T solution. The solution turns purple red. Titrate the solution with EDTA solution from the burette till the color of the solution changes from purple red to blue. Calculate the hardness of water. The hardness of water is expressed in parts per million units. 1 ml of 0.01M EDTA $\equiv$ 1 mg of $CaCO_3$.

Some indicators work on the competitive adsorption properties of the titrant and indicator with an adsorbant. For example iodine forms inclusion complex with starch which is colored. Iodine once is converted to iodide the color of starch reappears. This observation is used to determine the end point of an iodometric titration.

Certain titrations end point determination donot need use of an indicator. In such reactions one of the component changes its color during the course of reactions. The reactions like reduction of potassium permanganate to Mn^{2+}, reduction of potassium dichromate to Cr^{3+} etc. are such examples. In the former case a violet to colorless solution is observed, whereas in the later case orange to green coloration is observed.

A redox titration without indicator

Commercially available potassium permanganate generally contains impurity. It cannot be used as a primary standard. In order to make standard potassium permanganate solution it requires to be standardized by a primary standard. Generally potassium permanganate standardized by reacting it with oxalic acid in presence of a strong acid. The reaction is as follows

$$KMnO_4 + H_2C_2O_4 + H_2SO_4 \longrightarrow K_2SO_4 + MnSO_4 + CO_2$$

$$.....(2.86)$$

This titration does not require an indicator from external source as it is a self indicating system.

Standardization of oxalic acid

Fill a burette (50 ml) with potassium permanganate solution (0.01M approx). Pipette out standard oxalic acid (10 ml, 0.1M) into a beaker (250 ml). Add 10-15 ml of sulphuric acid (10N) to this and followed by distilled water (~75 ml); heat the solution to about 60°C. Titrate the solution with potassium permanganate solution. At first add potassium permanganate solution in small quantities with stirring; the pink color of potassium permanganate will take some time to discharge its color at the beginning. After some amount of potassium permanganate solution is added, the pink color will be discharged quickly. (This happens as it is an autocatalytic reaction, catalysed by manganese(II) ions). Add potassium permanganate solution drop wise with stirring until with one drop makes the whole solution pink. Note the volume of potassium permanganate solution added. Calculate the strength of the oxalic acid.

Calculation:

Let volume of $KMnO_4$ solution = V_1 ml, Strength of oxalic acid = S_1 (N)

Therefore, strength of $KMnO_4$ solution = $(10 \times S_1) / V_1$ (N)

Simultaneous estimation of more than one ions

It becomes essential to determine concentration of several metal ions as well as anions in a solution when they are present together. This is especially true when an ore is to be analyzed or purified. In such cases advantages of competitive reactions are taken. But the some other ions may interfere on estimation. So, in some case it becomes necessary to remove the interfering ions completely prior to the estimation. Consider the quantitative estimation of copper (II), calcium (II) and chloride ions in a mixture. To estimate these three ions advantage of the fact taken is the chloride ion can be separated by precipitation with silver nitrate and estimated. Whereas, between the Ca^{2+} and Cu^{2+}, Cu^{2+} is susceptible towards redox reaction with KI but Ca^{2+} is not. Thus, they can be estimated by independent experiments after adequate separation.

Copper (II) solution oxidizes potassium iodide and liberates iodine, liberated iodine is estimated by sodium thiosulphate. The ionic reactions involved in the estimation are:

$$2\,Cu^{2+} + 4I^- \Rightarrow 2Cu^{2+} + 2I_2 \qquad \qquad \dots\dots(2.87)$$

$$\text{and} \qquad 2I_2 + 2S_2O_3^{2-} \Rightarrow 2I^- + S_4O_6^{2-} \qquad \qquad \dots\dots(2.88)$$

The calcium is estimated by complexometric titration. But during the complexometric titration of calcium (II) and copper (II) can interfere, if eriochrome black-T indicator is used. So, copper (II) is reduced to copper (I) by using hydrazine hydrate and precipitated as copper thiocyanate by using ammonium thiocyanate and removed from the mixture. The calcium (II) ions is then titrated with ethylenediamine tetra-acetic acid (EDTA). The ionic equation involved is

$$Ca^{2+} + H_2Y^{2-} \Rightarrow CaY^{2-} + 2H^+ \qquad\qquad .....(2.89)$$

where, H_2Y^{2-} is dianion of EDTA

With calcium(II) ions alone, visual determination of end point is difficult as no sharp end point can be obtained with eriochrome black-T indicator. Addition of small amount of magnesium-ethylenediamine tetraacetic acid solution helps in determination of sharp end point.

Chloride ions can be estimated gravimetrically by precipitating it as silver chloride, the aqueous solution of chloride ion is acidified with dilute nitric acid, in order to prevent the precipitation of other silver salts.

$$Ag^+ + Cl^- \Rightarrow AgCl \qquad\qquad .....(2.90)$$

A slight excess of silver nitrate solution is added whereupon silver chloride is precipitated.

Estimation of copper(II)

Take 25 ml of stock solution containing mixture of ions in a conical flask and add potassium iodide to it. Titrate the liberated iodine with standard sodium thiosulphate solution.

Estimation of calcium(II)

Pipette out 25 ml of stock solution into a conical flask (250 ml). Add ammonium thiocyanate solid to this. A green coloration develops and then add hydrazinium sulfate solid. Heat the solution on a water bath. Remove the white precipitate formed by filtration. Take the filtrate in a conical flask and add ammonia-ammonium chloride buffer solution (pH=10) and add 1ml of magnesium-EDTA to the filtrate. Titrate the solution with standard ethylenediamine tetraacetic acid solution using eriochrome black-T as an indicator.

Estimation of chloride ions

Take 25 ml of stock solution in a conical flask. Acidify the solution with dilute nitric acid, and then add silver nitrate solution. This will give a white precipitate of silver chloride. Filter the precipitate in a pre-weighed sintered glass crucible and estimate the chloride gravimetrically.

Redox titrations

Ferrous iron can be estimated in solutions by volumetric titration. Titrating the ferrous iron present in a solution with potassium dichromate using diphenylamine indicator. The method enables one to estimate the amount of iron at +2 oxidation state present in the solution, without affecting the iron at +3 oxidation state. The redox reactions involved in the titration is as shown below

$$Cr_2O_7^{2-} + 14\ H^+ + 6\ e^- \rightarrow 2\ Cr^{3+} + 7\ H_2O$$

$$6Fe^{2+} \rightarrow 6Fe^{3+} + 6e-$$

$$Cr_2O_7^{2-} + 14\ H^+ + 6\ e^- + 7\ Fe^{2+} \rightarrow 2\ Cr^{3+} + 7\ H_2O + 6\ Fe^{3+} \quad .....(2.91)$$

The experiment can be extended to determine ferric ions as well as mixture of ferrous and ferric ions simultaneously. For such titrations the ferric ion in mixture is reduced to ferrous state with stannic chloride solution.

$$Sn^{2+} + 2\ Fe^{3+} \rightarrow 2\ Fe^{2+} + Sn^{4+} \quad .....(2.92)$$

Adding mercuric chloride oxidizes the excess stannous chloride in the solution. The mercurous chloride forms silky white precipitate of mercuric chloride on oxidation.

Procedure

Take iron solution (10 ml, about N/10) and add dilute sulphuric acid (15 ml) followed by one or two drops of diphenylamine indicator and phosphoric acid (2 ml, 1:1). Titrate the mixture with standard potassium dichromate solution (20 ml, N / 20). At the end point bluish-violet color will appear. This gives the amount of ferrous iron present in the solution.

To determine the ferric ion specifically in a mixture of ferrous and ferric ions in a solution, take 20 ml solution containing the ferrous and ferric iron (about N/10) and add concentrated hydrochloric acid (2 ml) to it. Boil the solution and add stannous chloride solution drop wise to the boiling mixture;

till the yellow color of the ferric ions disappears. Cool the solution and dilute it with water to 150 ml. Now add mercuric chloride solution (10 ml) to the mixture to oxidise the excess stannous chloride present in the solution. Titrate the solution with standard potassium dichromate solution (20ml, N/20) with diphenylamine as indicator. Calculate the total amount of ferrous and ferric iron present in the solution. Subtract the amount of ferrous ion. This will give the amount of ferric ion in the solution.

Potentiometric titrations

A redox couple present in a solution can be estimated by potentiometric titrations with another suitable redox couple using an electrochemical cell having platinum and calomel electrode. Calomel electrode serves the purpose of reference electrode. Consider a solution of stannous ions to which some ceric ions are added. The oxidation-reduction system contains excess of Sn^{2+} ions with some Sn^{4+} ions that are oxidized by ceric nitrate during the titration thereby variations occur in the electrode potential of Sn^{2+}/Sn^{4+} system. The e.m.f measurement of the cell can be used to determine the equivalence point of titration of Sn^{2+} against ceric ion as after the end point there will be new cell from Ce^{4+}/Ce^{3+} ions, as all the Sn^{2+} are consumed.

The electrode potential of Sn^{4+} / Sn^{2+} couple is given by the following equation

$$E = E_0 - \frac{RT}{nF} \log \frac{[Sn^{4+}]}{[Sn^{2+}]} \qquad \dots\dots(2.93)$$

where, E_0 is the standard redox potential of Sn^{4+} / Sn^{2+} system. This couple controls the e.m.f of the cell before the end point. Once the end point is reached the e.m.f. of the cell will be controlled by another equation based on Ce^{4+}/Ce^{3+} couple.

$$E = E_0 - \frac{RT}{nF} \log \frac{[Ce^{4+}]}{[Ce^{3+}]} \qquad \dots\dots(2.94)$$

Procedure

Take a solution containing Sn^{2+} ion (10 ml) and insert a platinum electrode on to it and connect it to the positive terminal of the potentiometer. Place a calomel electrode in saturated solution of potassium chloride and connect it to the negative terminal of the potentiometer. Put a potassium chloride salt bridge between the two electrolytic solution containing the electrodes. Once small amount of Ce^{4+} is agdded the cell may be represented by Sn^{2+}/Sn^{4+} couple, as some Sn^{2+} will get oxidized to Sn^{4+}

$$\text{Saturated calomel electrode} \parallel Sn^{2+}, Sn^{4+} \mid Pt$$

Measure the e.m.f of the cell. Add standard ceric nitrate solution (comparable strength of tin solution) in 1 ml portions to the ferrous solution at a time and measure the e.m.f of solutions after each addition. There will be a sharp increase in the e.m.f of the solution at the equivalence point due to taking over of the e.m.f by another cell namely

$$\text{Saturated calomel electrode} \parallel Ce^{3+}, Ce^{4+} \mid Pt$$

Plot a graph of e.m.f against volume of ceric nitrate solution added. Plot also the $\Delta E / \Delta V$ against volume of ceric nitrate solution. Determine the volume of ceric nitrate solution used to reach the equivalence point. Calculate the amount of tin (II) in the solution from this value.

Beer-Lambert Law

Spectroscopic techniques permit identification as well as quantitative estimation of one or more substrates simultaneously. For detection and estimation of multiple components independent signals that are well separated and non-superimposable are essential. UV-visible spectroscopy is a technique that is easy to perform and give very accurate results at low concentration limit. The Beer-Lambert law is a simple equation that gives the relationship between absorbance and concentration and is expressed as

$$A = \varepsilon\, c\, l \qquad\qquad \dots(2.95)$$

where A is absorption, l is the length of sample tube and c is the concentration of a solution under consideration. ε is a constant; which is characteristic of a sample. For quantitative estimation generally a calibration curve is made by measuring absorptions at a particular wavelength of various concentrations of the sample. From the calibration curve the absorption of unknown samples can be determined. Alternatively, if the molar extinction co-efficient and the path length are known the concentration can be directly calculated.

Visible spectroscopy for quantitative determination

Let us consider an experiment for simultaneous spectro-photometric determination of two solutes in a solution. The absorbances are additive, provided there is no reaction between the two solutes and solvents. Hence we may write

$$A\,\lambda_1 = \lambda_1\,A_1 + \lambda_1\,A_2 \qquad\qquad \dots..(2.96)$$

and $$\qquad A\,\lambda_2 = \lambda_2\,A_1 + \lambda_2\,A_2 \qquad\qquad \dots..(2.97)$$

where, A_1 and A_2 are the measured absorbances at the two wavelengths λ_1 and λ_2; the subscripts 1 and 2 refers to the two different substances, and the subscripts λ_1 and λ_2 refer to the different wavelengths. The wavelengths are selected to coincide with the absorption maxima of the two solutes; the absorption spectra of the two solutes should not overlap, so that substance 1 absorbs strongly at wavelength λ_1 and weakly at wavelength λ_2, and substance 2 absorbs strongly at λ_2 and weakly at λ_1. Now $A = \varepsilon\,c\,l$, where ε is the molar absorption coefficient is a constant at any particular wavelength, c is the concentration (mol L^{-1}) and l is the thickness, or length, of the absorbing solution (cm). If we set $l = 1$ cm then

$$A\,\lambda_1 = \lambda_1\,\varepsilon_1 c_1 + \lambda_1\,\varepsilon_2 c_2 \qquad\qquad \dots..(2.98)$$

and $$\qquad A\,\lambda_2 = \lambda_2\,\varepsilon_1 c_1 + \lambda_2\,\varepsilon_2 c_2 \qquad\qquad \dots..(2.99)$$

Solutions of the these simultaneous equations gives

$$c_1 = (\lambda_2\,\varepsilon_2\,A\,\lambda_1 - \lambda_1\,\varepsilon_2\,A\,\lambda_2) / (\lambda_1\,\varepsilon_1\,\lambda_2\varepsilon_2 - \lambda_1\,\varepsilon_2\,\lambda_2\,\varepsilon_1) \qquad \dots..(2.100)$$

$$c_2 = (\lambda_1\varepsilon_1\,A\,\lambda_2 - \lambda_2\,\varepsilon_1\,A\,\lambda_1) / (\lambda_1\,\varepsilon_1\,\lambda_2\varepsilon_2 - \lambda_1\,\varepsilon_2\,\lambda_2\,\varepsilon_1) \qquad \dots..(2.101)$$

The values of molar absorption coefficients ε_1 and ε_2 can be deduced from measurements of the absorbances of pure solutions of substances 1 and 2. By measuring absorbance of the mixture at wavelengths λ_1 and λ_2 , the concentrations of the two components can be calculated. For such purpose potassium permanganate and potassium dichromate may be considered as two solutions.

The compounds are intensely colored and have absorption spectra which have well separable absorption maximum. The visible spectra of two compounds dissolved in water are shown in Fig. 2.12. These two graphs are used to determine the two molar extinction coefficients ε_1 and ε_2,

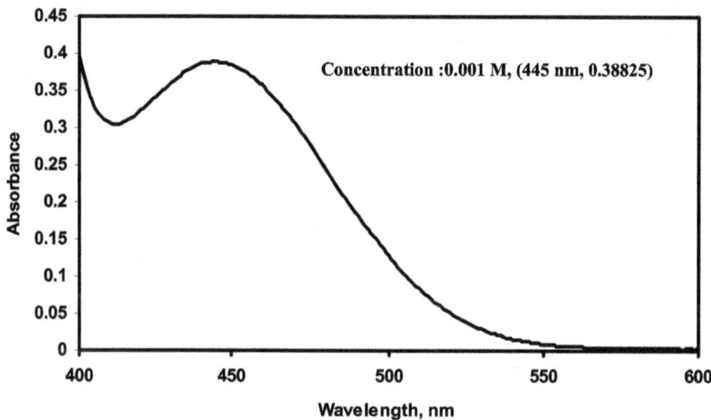

Potassium dichromate

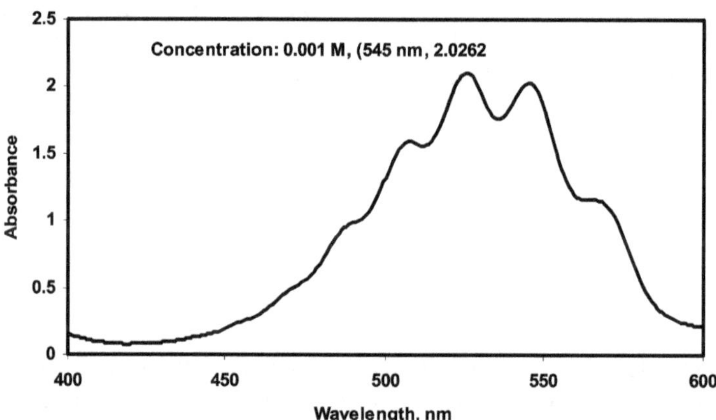

Potassium permanganate

Fig. 2.12 Visible spectra of aqueous solution of potassium dichromate (top) and potassium permanganate (bottom).

Procedure

Prepare a series of solutions, potassium dichromate 0.001M, 0.0005M and 0.00025M in mixed solvent of acids (1M sulphuric acid and 0.7M phosphoric acid). Prepare also a series of solutions of potassium permanganate 0.001M,

0.0005M, and 0.00025M respectively in mixed solvent of acids (1M sulphuric acid and 0.7M phosphoric acid). Measure the absorbances **A** for each of these solutions of potassium dichromate at 440 nm and for the solutions of potassium permanganate at 545 nm; by taking 1ml solution each. Calculate ε in each case by formula $A = \varepsilon cl$ and record the mean values for potassium dichromate (ε_1) and potassium permanganate (ε_2) at the two wavelengths. Mix 0.001 M potassium dichromate and 0.0005M potassium permanganate solutions in different known compositions to make 50 ml each (e.g. in ratios 1:2, 1:3, 2:3, 1:4, 3:2 etc) in 100 ml beakers. To each of these solutions add 1.0 ml of concentrated sulphuric acid. Measure the absorbance of each of the mixtures at 440 nm and at 545 nm. Calculate the absorbance of the mixtures from 2.102 and 2.103 and compare with the experimental results.

$$A_{440} = {}_{440}\varepsilon_{Cr} c_{Cr} + {}_{440}\varepsilon_{Mn} c_{Mn} \qquad .....(2.102)$$

$$A_{545} = {}_{545}\varepsilon_{Cr} c_{Cr} + {}_{545}\varepsilon_{Mn} c_{Mn} \qquad .....(2.103)$$

Determination of composition of a complex

The UV-visible spectroscopy can be used to determine composition of a complex.

Consider the following equilibrium

$$Fe^{3+} + x\ SCN^- \rightleftharpoons [Fe(SCN)_x]^{3-x} \qquad .....(2.104)$$

In this equilibrium an unknown number of x, is the number of thiocyanate ions, to form a coordination complex with the iron(III) ions. The number x is determined by experiment based on visible spectroscopy.

From a number of solutions containing different amounts of solutions of ferric ion and thiocyanate anion of known concentration, the number of moles of ferric ion and thiocyanate anion added to make the solution may be calculated from the absorbances of each solution. Let n_{SCN} and n_{Fe} be the number of moles of thiocyanate anion and ferric ion added to make a solution. From the calculated number of moles of ferric ion and thiocyanate anion, the mole fraction of thiocyanate anion in the mixture can be determined as follows:

$$\chi_{SCN} = n_{SCN}/(n_{Fe} + n_{SCN}) \qquad .....(2.105)$$

In the mixture that contains stoichiometric amounts of ferric ion and thiocyanate anion, according to equation 2.105, that:

$$x(n_{Fe}) = n_{SCN} \qquad \qquad .....(2.106)$$

Dividing both sides of equation 2.106 by the quantity $n_{Fe} + n_{SCN}$:

$$x\left(\frac{n_{Fe}}{n_{Fe} + n_{SCN}}\right) = \frac{n_{SCN}}{n_{Fe} + n_{SCN}} \qquad .....(2.107)$$

The right-hand side of equation 2.107 is equal to the mole fraction of thiocyanate anion, χ_{SCN}. The quantity in the parentheses on the left-hand side of equation 2.107 is equal to the mole fraction of Fe, χ_{Fe}. Making these replacements in equation 2.107 one obtains:

$$x\,\chi_{Fe} = \chi_{SCN} \qquad \qquad .....(2.108)$$

Since the sum of mole fractions in a mixture always add up to one, the mole fraction of iron (III) is related to the mole fraction of thiocyanate anion by the following equation:

$$\chi_{Fe} = 1 - \chi_{SCN} \qquad \qquad .....(2.109)$$

Substituting equation 2.108 into equation 2.109, the following relationship is obtained:

$$x\,(1 - \chi_{SCN}) = \chi_{SCN} \qquad \qquad .....(2.110)$$
$$x = \chi_{SCN} /(1 - \chi_{SCN}) \qquad \qquad .....(2.111)$$

Therefore, for the mixture containing a stoichiometric amount of iron (III) and thiocyanate anion, the coordination number, x, may be calculated from the value of the mole fraction of thiocyanate anion in the mixture.

Procedure

Prepare a series of ten numbers of thiocyanate solutions with ferric nitrate by varying from 0-1 mole fractions with an increment of 0.1 mole fraction. Record the visible absorbance in the range 350-600 nm and find out the λ_{max} for absorption by taking one of the solution. λ_{max} will be at 430 nm. From each of these solutions take 1.5 ml solution each to glass cuvette and measure absorbance at this λ_{max}. Plot absorbance versus mole fraction of thiocyanate anion and then find the mole fraction at which the maximum absorbance occurs. This mole fraction of thiocyanate anion, corresponds to the number of moles of iron(III) and thiocyanate anion needed to form the complex. Find out the value of x from expression 2.111.

Optical rotation

Optically active substances are those compounds, which can rotate the plane of polarized light. Optical activity of a compound is determined by optical rotation. The compounds, which rotate the plane of polarized light to the right, are called dextrorotatory and those which rotate to the left, are called laevorotatory. In the case of a solution containing chiral compound/s the angle of rotation depends on (a) nature of the substance, (b) length through which light travels, (c) wavelength of the light, and (d) temperature.

The specific rotation for an optically active substance in solution is defined as the angle of rotation caused by the solution containing one gm of the active substance in one ml of the solution when the length of the column of solution through which the light passes is one decimeter. The value of specific rotation depends on the temperature and wavelength of the light used.

Mathematically, specific rotation of an optically active compound is defined as

$$[\infty]^t_\lambda \;=\; 100\,\infty/lc \qquad\qquad .....(2.112)$$

Where,

$[\infty]^t_\lambda$ = The specific rotation at temperature t $^\circ$C and for light of wavelength λ

∞ − Observed angle of rotation

l = Length of column of solution in dm

c = No. of grams of the active substance in 100 ml of the solution.

From this equation it is clear that by determining specific rotation of a optically active compound such as sucrose, the concentration of a sucrose solution can be determined.

Procedure

Take five volumetric flasks (25 ml) and prepare 2%, 4%, 6%, 8% and 10% aqueous solution of sucrose in water. Fill the polarimeter tube with any of the solution and measure the optical rotation. Calculate the specific rotation in each case. Plot the specific rotation vs the corresponding concentration in (gm / 100 ml). A straight line will be obtained. From the plot, find out the concentration of the unknown sucrose solution.

The optical rotation can be used to determine rate constant of a reaction in which the optical rotation changes with time.

Enzyme activity

Enzymes are biological catalysts, their reactivity depends on the protein associated with them. The structures of enzymes are governed by many factors such as pH, ionic strength, and temperature. Small changes in these parameters can denature a protein structure. Tyrosinase is an enzyme found in potato and many plants and animals. Tyrosinase is the casuse of darkening of skin on exposure to sunlight and also darkening of potato or banana, apple, or pear on exposure to air. It reacts with a variety of natural substrates containing tyrosine, catechol (Fig. 2.13) and other dihydroxy aromatics.

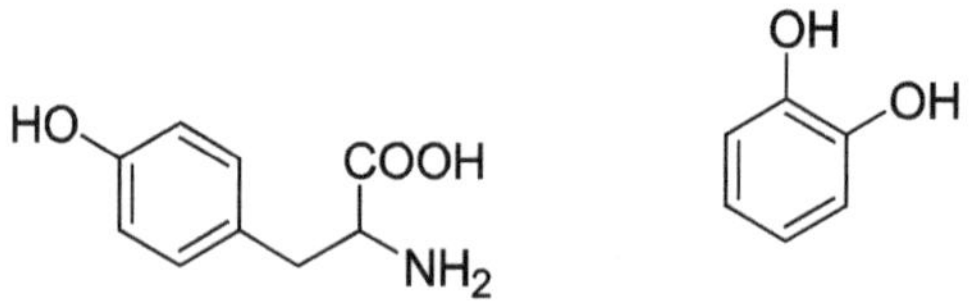

Tyrosine

Catechol

Fig. 2.13

Let us consider the oxidation of catechol which is catalysed by tyrosinase enzyme:

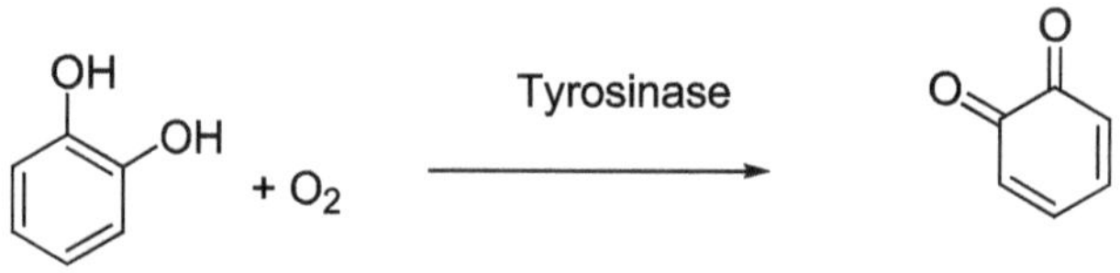

$$.....(2.113)$$

Tyrosinase extracted from potato can be used to study this catalytic reactiion. It is studied by reacting catechol with air in the presence of catalytic amount of tyrosinase. The concentration of the ortho-quinone formed in the solution is monitored by visible spectroscopy.

Procedure for enzyme kinetics

Take some peeled potato (150-200 gm), wash and chop them into small pieces and keep for about one minute in ice-cold distilled water (40 ml).

Filter the homogenenised mixture through layers of cheese cloth and transfer to a centrifuge tube. Centrifuge it for five minutes and dilute the supernatant liquid with ice-cold distilled water. Store the supernatant liquid in a freeze. The supernatant liquid contains tyrosinase enzyme. Make a solution of catechol (0.006M, 5 ml), sodium phosphate buffer (0.1M, pH 6.0, 5 ml) in distilled water (15 ml at room temperature). Mix 2 ml each of the enzyme and the catechol solutions. Monitor the reaction by measuring the absorbance of the solution at 540 nm at definite time interval (half minutes). Plot the absorbance vs time. A linear graph will be obtained. The reactions can be monitored at different temperature and pH to find out the effect of temperature, pH and also to determine the activation energy.

Oscillatory reactions

There are some chemical reactions that causes color change in cyclic manner with time and the reappearance and disappearances of color takes place from time to time. Such reactions are referred to as chemical clock.

Let us consider a reaction in which A converts to B, and B converts to C, and C converts to D.

$$A \rightleftharpoons B \rightleftharpoons C \rightleftharpoons D$$

There are three steps in this sequence of reactions; and each of the step is associated with equilibrium constants. Depending on the relative rate of each backward and forward step, the reactions will follow different rate equations for each step and overall rate of formation of product will depend on these steps.

We can further complicate the reaction by introducing some more steps in which the intermediates can react with the product so that reaction follows the following steps

$$A \xrightarrow{k_1} B$$
$$B \xrightarrow{k_2} C$$
$$B + 2C \xrightarrow{k_3} 3C$$
$$C \xrightarrow{k_4} D$$

Scheme 2.7 A representation of an oscillatory reaction

Thus, rate of formation of D will be highly decided by how much B and C are produced. Such a situation occurs in the reaction of sodium iodate with H_3AsO_3. The intermediate steps in the reaction can be written as follows:

$$NaIO_3 \longrightarrow Na^+ + IO_3^-$$

$$IO_3^- + AsO_3^{3-} \longrightarrow I^- + AsO_4^{3-}$$

$$IO_3^- + I^- + AsO_3^{3-} \longrightarrow I^- + AsO_4^{3-}$$

$$I^- \longrightarrow I_2$$

Scheme 2.8 Iodine clock reaction

In these reactions as iodine concentration increases the concentration of iodate will be decided by iodide ions, but the source of iodide is iodate. Thus, the reaction will oscillate back and forth depending on the amount of H_3AsO_3 used in the reaction. These reactions are known as oscillatory reactions. Similarly, if the same reaction is carried out with sulphurous acid, it will follow the following steps:

$$IO_3^- + HSO_3^- \rightarrow I^- + HSO_4^-$$

$$IO_3^- + I^- + 6H^+ \rightarrow I_2 + H_2O$$

$$I_2 + HSO_3^- + H_2O \rightarrow I^- + HSO_4^- + H^+$$

Scheme 2.9 Iodine clock reaction by iodate and sulphurous acid

First step of these three steps is the rate determining step. The excess iodate oxidizes iodide to iodine. But, the amount of iodine is reduced again immediately by its conversion to iodide by the bisulfite. The formation of iodine will be there when all bisulphite is consumed. Thus, by adding bisulphite time to time the color reappearance of iodine will take place.

Similar reaction is also observed during the reaction of potassium iodide with hydrogen peroxide. In this reaction first step is formation of triodide anion

$$H_2O_2 + I^- + H^+ \rightarrow I_3^- + H_2O \qquad \dots\dots(2.114)$$

The triodide formed can be reacted with sodium thoisulphate to give iodide again.

$$I_3^- + S_2O_3^{2-} \rightarrow I^- + S_4O_6^{2-}$$ .....(2.115)

Thus, after some time the solution develops color due to formation of iodine. The process can reoccur by adjusting the concentration of reagent and can be set to colorise and decolorise at specific time, thereby it can act like clock.

Out look

It is clear that the principles involved in each experiment not only helps to understand the chemical process around it; but also helps in making new reaction schemes for development of new useful products. For example, calcium and carbon reacts together at high temperature to form calcium carbide. Calcium carbide on reaction with water liberates acetylene gas; the acetylene gas is commercially used for welding purpose. Certain processes may look trivial, but underneath it may have important utility. Let us look at the hydrolysis of magnesium nitride, this reaction leads to ammonia and magnesium hydroxide. But the same reaction carried out in deuterium oxide leads to pure deuterated ammonia which is otherwise difficult to make in pure form.

$$Mg_3N_2 + H_2O \longrightarrow Mg(OH)_2 + NH_3$$

$$Mg_3N_2 + D_2O \longrightarrow Mg(OD)_2 + ND_3$$

Nitrogen is a common element found as gas in nature. The purification of nitrogen from air and liquefaction of nitrogen makes understanding of physical properties of nitrogen. Liquid nitrogen a versatile inert liquid used to generate low temperature. Nitrogen not only is a resource for ammonia, which also serves as precursor for urea, nitric acid, hydrazine etc. Urea has applications in inclusion compound formation, as fertilizer and as source for polymeric materials such as melamine, phenol-formaldehyde resin. Hydrazine is a rocket fuel, highly useful reducing agent in organic chemistry. This suggests that series of interrelated industrial products uses nitrogen as resource. However, small a reaction scheme is, the essential part of it is the formation of it as basis of study to explore and design a useful material. To achieve this economically and efficiently, understanding from molecular level becomes essential.

Chapter **3**

Experimental Coordination Chemistry

Coordination Complex

Coordination chemistry deals with metal complexes with focus on metal centers and ligands. The complex ions of coordination complexes are formed by the reaction of a metal ion with Lewis base/s, which are known as ligand/s. The metal complexes are associated with different coordination numbers. The coordination sites in a metal complex are occupied by ligands in the coordination sphere. Depending on the charge on the metal ion of the complexes, they can be neutral, anionic, or cationic. The ligands may also be anionic, cationic or neutral. The charges on the complex ion are finally decided by the charges of the metal ions as well as the charge of the ligands. In this chapter, we shall take examples of some metal complexes and discuss about their properties and coordination behaviors. The coordination complexes may be formed from two or more independent chemical species having independent existence. On many occasions, bindings of the metal ion with ligands are reversible. Thus, equilibrium between the solvated metal ion and the ligand exists.

For example, when nickel(II) chloride is reacted with ammonium hydroxide solution it forms the hexaamminenickel(II) chloride. The reaction can easily be reversed by adding hydrochloric acid to give back the starting materials. The complexation may lead to interesting color changes; which becomes fascinating to a beginner. For such an experiment, take about 500 ml water in a beaker (1000 ml). Add nickel nitrate solution to it (40 ml, 1M). The solution turns blue. This happens because it forms hexaaquanickel(II) nitrate. The color of the solution will become purple on addion of ethylenediamine solution (40ml, 25%), to this solution. This is due to the formation of *tris*-ethylenediamminenickel(II) nitrate. Further addition of dimethylglyoxime (abbreviated as dmgH) (50 ml, 1 % w/v) to this solution makes the solution pink; the formation of *bis*-dimethylglyoximatonickel(II) complex on addition of dmgH leads to pink coloration; and finally addition of potassium cyanide (200 ml, 2M) to the pink solution makes the solution orange. This happens due to formation of potassium hexacyanonickelate(II). In all these cases different hexa-coordinated nickel (II) complexes are formed. In these complexes the central metal ion is same. The reactivities of each of the ligand used here are different; so ligand substitution is possible in step wise manner, which results in the different colors. Each of the complex has characteristic electronic features; the first two complexes formed from addition of ammonia solution and ethylenenediamine are cationic, whereas the third one is neutral, and the last one when cyanide is the ligand it is anionic. The strength of the ligands are increasing in the order H_2O < en < dmg < CN^- (en = Ethylenediamine, dmg = Dimethylglyoximato anion). Depending on the ligands the crystal field will be different and electron pairing will be favored by the cyanide ligands which is strongest among the four ligands under consideration. Thus, we may say that the complexes are varying from high spin to low spin. Each of these reaction is associated with an equilibrium constant, which decides favorability of the processes. Following reaction sequences takes place during subsequent addition of different ligands (solutions) to the aqueous solution of nickel(II) nitrate.

$$[Ni(H_2O)_6](NO_3)_2 + 6\ NH_3 \rightarrow [Ni(NH_3)_6]\ (NO_3)_2 + H_2O$$

$$K = 1.2 \times 10^9 \qquad\qquad\qquad .....(3.1)$$

$$[Ni(NH_3)_6](NO_3)_2 + 3\ en \rightarrow [Ni(en)_3]\ (NO_3)_2 + 6NH_3$$

$$K = 1.1 \times 10^9 \qquad\qquad\qquad .....(3.2)$$

$$[Ni(en)_3](NO_3)_2 + 2\ dmgH \rightarrow [Ni(dmg)_2] + 3en + HNO_3$$

$$[Ni(dmg)_2] + 4\ KCN \rightarrow K_2[Ni(CN)_4] + 2\ Kdmg$$

$$K = 6.3 \times 10^7 \qquad\qquad\qquad .....(3.4)$$

To understand about such equilibrium processes, in another experiment make a solution of cobalt(II) chloride (about 3 g) in ethanol (100 ml) in a beaker (250 ml). The color of the solution is pink. It is because of hexaaquacobalt(II) chloride. Add about 1ml of concentrated hydrochloric acid, the solution turns purple. Take about 40 ml solution in a conical flask. Heat the solution, a blue color will appear. Cool the solution pink color reappears. At room temperature the solution will have an intermittent color. It occurs due to the equilibrium shown in equation 3.5.

$$[CoCl_4]^{2-} \text{ (blue)} + 6H_2O \rightleftharpoons [Co(H_2O)_6]^{2+} \text{ (pink)} + 4Cl^- \text{ (aq)}$$

$$.....(3.5)$$

This suggests that equilibrium between two complexes in solution can be brought about by temperature. The equilibrium may be pushed towards right or left side of a balanced chemical equation by adding adequate ligand.

Dissolution of copper(II) sulphate in ammonium hydroxide is a common observation for demonstration of complex formation in solution. The complex tetraamminecopper(II) hydroxide formed in this reaction is used for preparation of Rayon from cellulose. A solution of tetraamminecopper(II) hydroxide can be prepared by adding concentrated ammonium hydroxide drop by drop to an aqueous solution of copper(II) sulphate pentahydrate until a light blue color and a precipitate is formed. The precipitate is the tetraamminecopper(II) hydroxide. If the solution becomes dark blue, it leads to the formation of hexaamminecopper(II) hydroxide. Commonly used laboratory filter paper is comprised of cellulose; it can be converted to another polymeric form Rayon. To prepare Rayon from filter paper, tear a piece of filter paper into small pieces and place the pieces in a small beaker. Add a solution of tetraamminecopper(II) hydroxide to it; the filter paper will get completely dissolved. Fill a dropper or small syringe with the dissolved paper. Using gentle pressure on the syringe, allow the solution from the syringe to pass into a beaker containing dilute sulphuric acid. This leads to the formation of the white rayon fibers. In this process the cellulose in the filter paper gets hydrolysed. The hydrolysis caused by tetraamminecopper(II) hydroxide leads to product soluble in water. When the solution is acidified, cellulose is regenerated in the form of hard filaments, which is called Rayon.

The coordination number of a complex varies with the oxidation state. For example the reduction of pentaamminecopper(II) hydroxide with sodium dithionate reduces copper(II) to copper(I), producing the diamminecopper(I) hydroxide; which is a colorless copper(I) complex. This is an example of a transformation of a five-coordinated complex to two coordinated complex due to change of oxidation state of the metal. If excess dithionite is not used in this reaction, the solution becomes blue at the air-water meniscus. Slowly the entire reaction mixture turns blue due to air oxidation of the copper(I) back to copper(II). Addition of dithionate to such a solution makes it colorless. But, the addition of sulphuric acid to the solution makes the solution acidic and it also leads to precipitation of reddish copper metal powder. During this process, the solution becomes opaque reddish brown and turns blue. This color change happens due to the disproportionation reaction of copper(I) to copper(II) and copper(0) as illustrated in equation 3.6.

$$2Cu(I) \rightarrow Cu(0) + Cu(II) \qquad\qquad .....(3.6)$$

Copper(II) complex tetramminecopper(II) sulphate hydrate, with four coordination number can be prepared from anhydrous white copper(II) sulphate in water by reaction with ammonia.

There are also examples of some reactions that leads to precipitation of copper(I) complexes or salts. Such insoluble salts or complexes do not disproportionate easily. For example, the reaction of copper(II) sulphate with potassium iodide liberates iodine and copper(I) iodide is formed in this reaction (equation 3.7). Copper(I) iodide can be filtered from the reaction mixture and stored in   desiccator. Copper(I) iodide is stable under ordinary condition.

$$CuSO_4 + KI \longrightarrow I_2 + CuI\downarrow + K_2SO_4 \qquad .....(3.7)$$

The reduction of copper(II) perchlorate hexahydrate by copper(0) metallic powder in acetonitrile solvent leads to formation of a copper(I) complex. In this reaction the solvent acetonitrile plays role of ligands also. The copper(I) complex formed in this reaction is tetraacetonitilecopper(I) perchlorate. This less stable oxidation state +1 of copper in this complex is stabilized by the acetonitrile ligands. The

$$Cu(ClO_4)_2\,6H_2O + Cu \xrightarrow[\text{CH}_3\text{CN}]{\text{Reflux}} \left[\begin{array}{c} CH_3CN \\ CH_3CN-Cu-NCCH_3 \\ H_3CCN \end{array} \right] ClO_4$$

Colourless solid

$$.....(3.8)$$

acetonitrile has ability to stablise low oxidation state of metal, in this case copper(I) by back bonding to the metal. By virtue of this back bond formation the extra electron density at metal site due to low oxidation state of a metal is reduced to make a synergic effect for making a stable complex. The perchlorate anion is outside the coordination sphere and the complex is ionic in nature. The ionic nature of the perchlorate can be easily confirmed by its strong IR absorptions around $1100\ cm^{-1}$.

Nomenclature of complexes

Based on the identity of the metal, its oxidation state, the number and type/s of ligand/s, and the identities of the other cations and anions present in a complex, International Union of Pure and Applied Chemistry (IUPAC), nomenclature is used to assign names to compounds. Some examples are $[Pt(NH_3)_2Cl_2]$, diamminedichloroplatinum(II); $[Co(NH_3)_5Cl]Cl_2$ pentaamminecobalt(III) chloride; $K_4Fe(CN)_6]$, potassium hexacyanoferrate(II) etc.

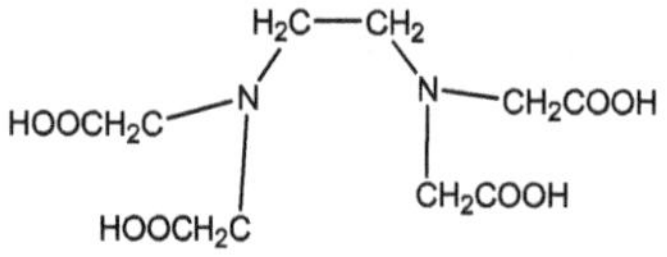

Ethylenediaminetetraaceetic acid or
1,2-ethanedinyldinitrilo-N,N'-
tetraacetatic acid

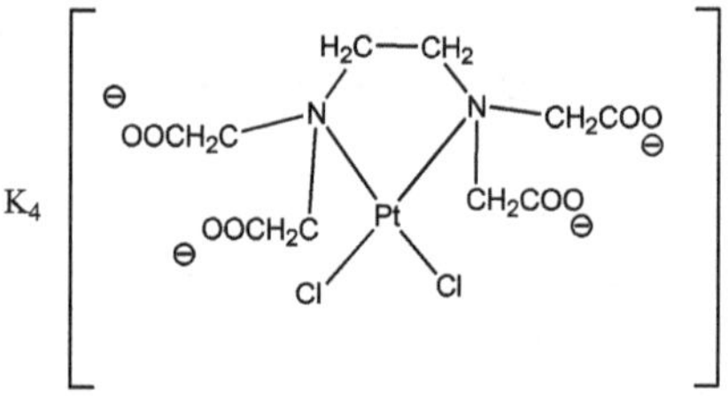

Pottasium dichloro(1,2-ethanedinyldinitrilo-κ^2N,N')
tetraacetato]platinate

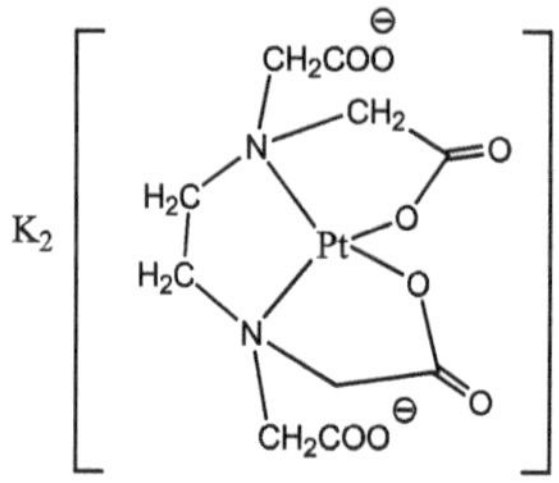

pottasium (1,2-ethanedinyldinitrilo-
κ^2N,N')tetraacetato-κ^2O,O']
platinate

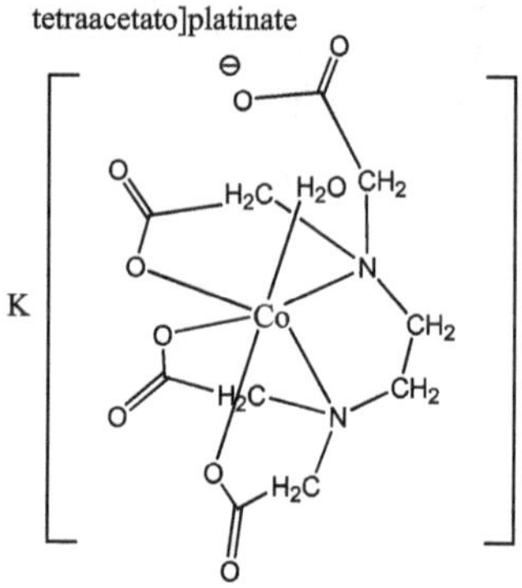

Pottasium aqua[(1,2-
ethanedinyldinitrilo-
κ^2N,N')(tetraacetato-κ^3O,O'', O''']
cobaltate

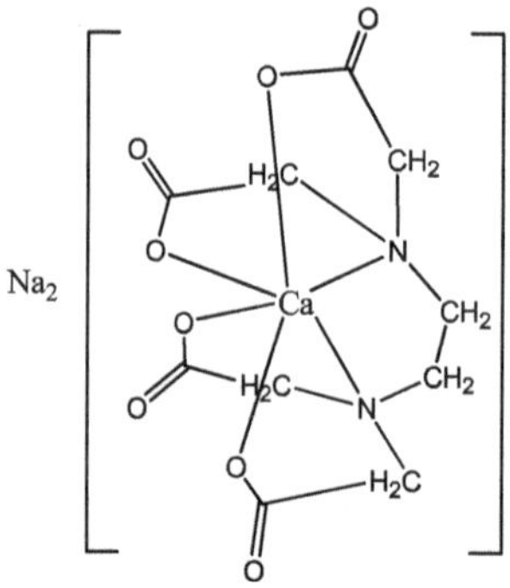

Sodium [(1,2-ethanedinyldinitrilo-κ^2N,N')(tetraacetato-κ^3O,O'', O''', O'''']calciate

Fig. 3.1 Name and structure of ethylenediamine tetraacetate complexes

Multi-dentate ligands such as anions of ethylenediaminetetraacetic acid can complex to a metal in different ways by binding at multiple binding sites. Few complexes of the ligand are shown in the Fig. 3.1 with their IUPAC name. The discussion on complete naming of inorganic compounds is beyond the scope of this chapter, however, the rules for common naming are available in any standard inorganic chemistry text books.

Hexaamminecobalt(III) chloride

Let us consider the complex hexaamminecobalt(III) chloride. It can be prepared by reacting cobalt(II) chloride with ammonium chloride and ammoium hydroxide followed by aerial oxidation or oxidation by hydrogen peroxide. The later step is to oxidize the cobalt +2 to cobalt +3 oxidation state.

$$2CoCl_2.6H_2O + 2NH_4Cl + 10NH_3 + H_2O_2 \longrightarrow$$

$$2[Co(NH_3)_6]Cl_3 + 2H_2O \quad .....(3.9)$$

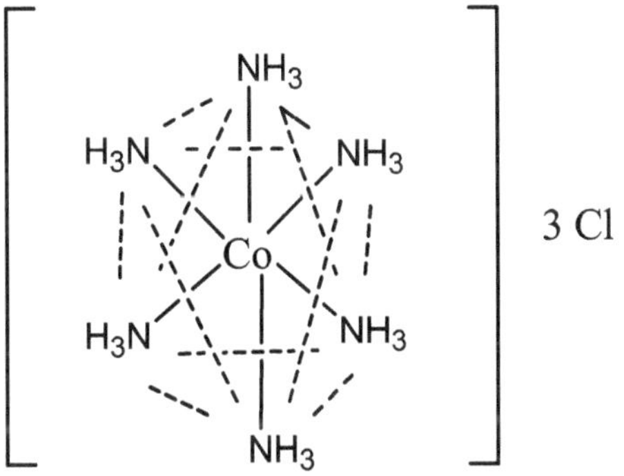

Fig. 3.2 Structure of hexaamminecobalt(III) chloride

The cobalt in this complex is in +3 oxidation state. It has six ammonia ligands coordinated to cobalt(III) ion and has an octahedral geometry. The composition of the complex can be deduced by estimating chloride and cobalt ions in the complex. The cobalt in oxidation state +3 is highly oxidizing agent and it gets converted to +2 oxidation state in the presence of a reductant. It readily reacts with potassium iodide to liberate iodine. The liberated iodine from such reaction can be estimated by titration with sodium thiosulphate. By this way, the amount of cobalt in the complex can be estimated. For such estimation the hexaamminecobalt(III) chloride is first decomposed by sulphuric acid. The amount of cobalt 3+ ions can also be determined by atomic absorption spectroscopy. The chloride ion in the complex is can be determined gravimetric method. For this purpose the chloride ions are precipitated as silver chloride by treating with silver nitrate. From the weight of the silver chloride obtained in the form of precipitate from such reactions the percentage of chloride in the complex is estimated. Based on the amount of the cobalt(III) cations and chloride anions, one can propose the structure. However, the nitrogen, hydrogen and halogen content of the complex can be alternatively determined by equipment called elemental analyzer also.

The complex shows magnetic moment close to zero Bohr Magneton (BM). This shows the complex to be a diamagnetic complex. The conductance measurement of the complex shows its molar conductance characteristic of 1:3 electrolyte.

The infra-red spectra of the complex shows symmetric NH-stretching at 3240 cm^{-1} and 3160 cm^{-1}, degenerate asymmetric stretching at 1615 cm^{-1} and symmetric ammonia angle deformation appears at 1308 cm^{-1}. Free ammonia has these absorbances at 3350 cm^{-1}, 3419 cm^{-1}, 1628 cm^{-1} and 950 cm^{-1} respectively.

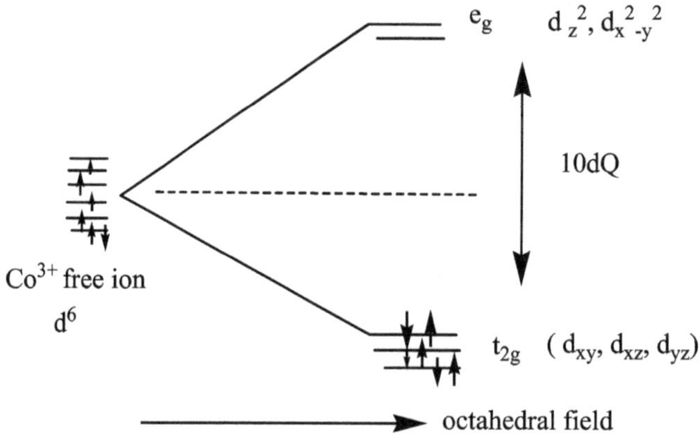

Crystal field splitting of $[Co(NH_3)_6]Cl_3$ showing it to be diamagnetic

Fig. 3.3

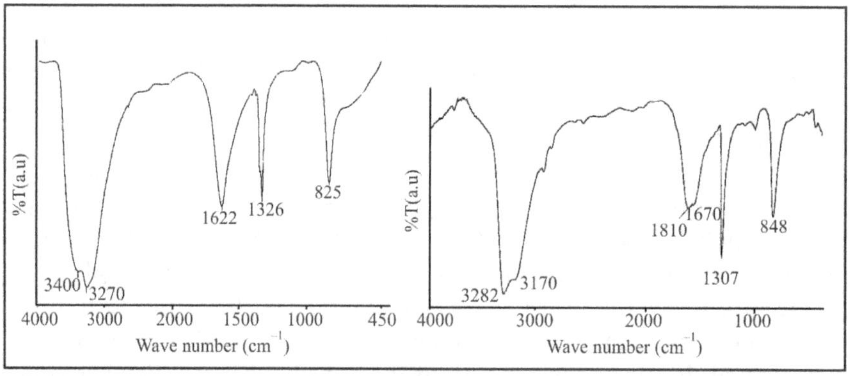

Hexaamminecobalt(III) chloride   Pentaamminechlorocobalt(III) chloride

Fig. 3.4 FT-IR spectra (KBr, cm^{-1})

Synthetic procedure for hexaamminecobalt(III) chloride

Make a paste of ammonium chloride (1.2 g, 2.2 mmol) by dissolving it in water (3 ml). Heat the paste till it starts boiling. Add cobalt(II) chloride hexahydrate (1.8 g, 7.5 mmol) to the paste. Take activated charcoal (0.2 g) in a separate conical flask and add the paste to this. Cool the flask and its contents under running water. Add concentrated ammonia (4.5 ml, 20% v/v) to the solution and cool the mixture to 10^0 C (approximately). Add slowly hydrogen peroxide (2.4 ml, 30 % v/v) to this solution with a dropper. Swirl the flask gently during the addition. Heat the solution on a steam bath at 50-60 °C until the pinkish tint disappears from the solution (~20 min). Cool the solution in an ice bath and filter it through a Buchner funnel. Transfer the residue to a boiling solution of hydrochloric acid (15 ml, 0.5 M). Heat the solution with stirring until the solution boils and filter the hot solution by using a Buchner funnel under suction. Collect the hexaamminecobalt(III) chloride as the filtrate and evaporation results the desired product, dry the product at 110 °C for an hr.

Pentaamminechlorocobalt(III) chloride

Pentaamminechlorocobalt(III) chloride can be prepared by a similar procedure to the hexaamminecobalt(III) chloride but with an additional treatment with concentrated hydrochloric acid.

$$CoCl_2 + NH_4Cl + NH_3 + H_2O_2 \rightarrow [Co(NH_3)_6]Cl_3 + HCl$$
$$\rightarrow 2[Co(NH_3)_5Cl]Cl_2 \qquad .....(3.10)$$

The complex has five ammonia ligands and one chloride ligand. It has a distorted octahedral geometry. The complex has two types of chloride ligands. One chloride anion act as a ligand and is positioned inside the coordination sphere. This chloride ligand does not ionise. It has also two ionic chlorides, which ionises in solution.

The IR spectrum of the complex shows symmetric NH-stretching at 3290 cm^{-1} and 3175 cm^{-1}, degenerate asymmetric stretching appears at 1589 cm^{-1} and symmetric NH_3 angle deformation appears at 1310 cm^{-1}.

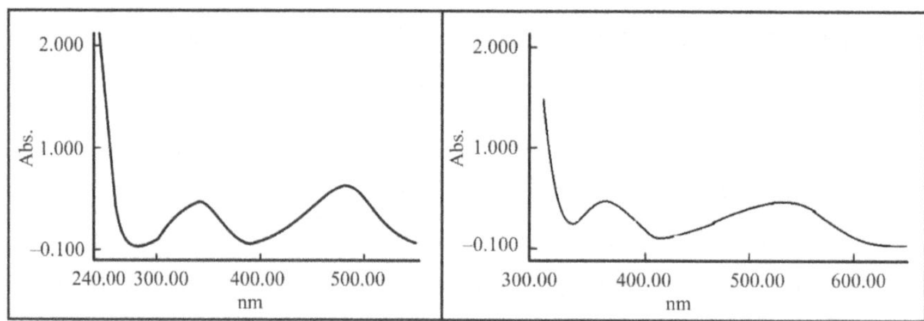

Hexaamminecobalt(III) chloride   Pentaamminechlorocobalt(III) chloride

Fig. 3.5 UV-visible spectra in water

It is interesting to compare the electronic spectra of the two complexes namely hexaamminecobalt(III) chloride with that of pentaamminechlorocobalt(III) chloride. In both these cases the metal ions are at +3 oxidation state. The cobalt at +3 oxidation state has a d^6-electronic configuration. The spin allowed transitions for low spin d^6-electronic are $^1A_{1g} \rightarrow {}^1T_{1g}$ and $^1A_{1g} \rightarrow {}^1T_{2g}$. In the case of hexaamminecobalt(III) chloride these transitions are observed at 340 nm and 475 nm as two distinctly distinguishable absorption peaks. In the case of pentaamminechlorocobalt(III) chloride there are two closely spaced absorptions at 365 nm and another absorption at 530 nm. Since the center of inversion is not present in the pentaamminechlorocobalt(III) chloride, the extinction coefficient of these transitions are slightly higher

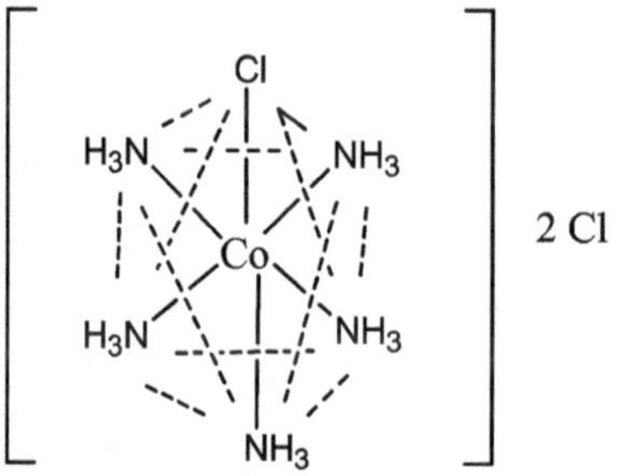

Fig. 3.6 The structure of pentaamminechlorocobalt(III) chloride

than that of hexaamminecobalt(III) chloride. In the case of pentaamminechlorocobalt(III) chloride asymmetry in the structure is caused by replacing one of the ammonia ligand by a chloride ligand. If the removal of the ammonia has taken place from $d_z{}^2$ orbital, $d_z{}^2$ orbital will be under different environment from the other orbitals. Thus, the relative energy of d_z^2, d_{xz} and d_{yz} will be affected by this unsymmetry caused to the octahedral structure. The stronger π- interaction of chloride raises the energy of d_{xz} and d_{yz} and a weak sigma interaction from chloride ion lowers the energy of d_z^2 orbital. So the degeneracy of triplet state takes place and the $^1T_1$ state splits into E_1 and A_1 states. Such a situation changes the electronic states from that of an octahedral geometry and results in the modified electronic transitions; namely $E_1 \rightarrow A_1$, $E_1 \rightarrow T_1$ and $A_1 \rightarrow T_1$. The former two transitions appear at 365 nm whereas the last one occurs at 530 nm.

Synthetic procedure for pentaamminechlorocobalt(III) chloride

Add a slurry prepared from cobalt(II) chloride hexahydrate (1.8 g) in water (2.5 ml) to a solution of ammonium chloride (5.6 g) in concentrated ammonium hydroxide (5.6 ml). Add very slowly (two drops at a time with caution) hydrogen peroxide (1.1 ml, 30%) with a dropper to this mixture. While adding, stir the solution vigorously. Heat the resulting mixture on a steam bath until thick paste is obtained (alternately the oxidation can be done by purging oxygen/air through solution for 10 hrs at room temperature). If the evaporation is done rapidlyt or the hydrogen peroxide is added fast, the paste shows a blue tint. In such a case, the experiment has to be repeated. Otherwise add hydrochloric acid (3 M, 22.5 ml) to the paste and heat at 60 °C for 10 min. Cool the solution to room temperature and filter in a Buchner funnel with suction. Wash the product with three 2 ml portions of ice water and then with three 2 ml portions acetone and dry under suction.

Transfer the product along with the filter paper to a solution of ammonia (2 M, 68 ml). Warm the resulting solution on a steam bath at 60 °C. Once the entire solid dissolves, filter the hot solution with suction. Reheat the filtrate on a steam bath with stirring. Add hydrochloric acid (12 M) in three times portions (17 ml each). Cool the solution to room temperature and filter. Wash the product with water (2 ml) three times followed by three times with acetone (2 ml).

Ethylenediamine as ligand

Ammonia is a monodentate ligand, whereas ethylenediamine is a bidentate ligand. The ethylenediamine forms dative bond with metal ions to make five-member ring structures. Such cyclic binding mode of bidentate ligand is called chelation and the complex formed through chelation is called chelate complex. The chelate complex formation is more thermodynamically favorable. When there are alternative possibilities of formation of bridging structure and chelate structure, the chelation over bridging compounds prevails when provision for five-member ring with the metal and ligand is present. The chelate formation depends on the ring strain and related steric factors. Using ethylenediamine as an ligand different metal complexes are prepared with different stoichiometry. The cobalt(III) ions prefer to have six-coordination geometry and it can form varieties of complexes with ethylenediamine. A simple example of a complex between cobalt(III) ion with ethylenediamine is dichloro-*bis*(ethylenediamine)cobalt(III) chloride

$$NH_3 \qquad\qquad NH_2CH_2CH_2NH_2$$

Monodentate ligand   bidentate ligand

Fig. 3.7

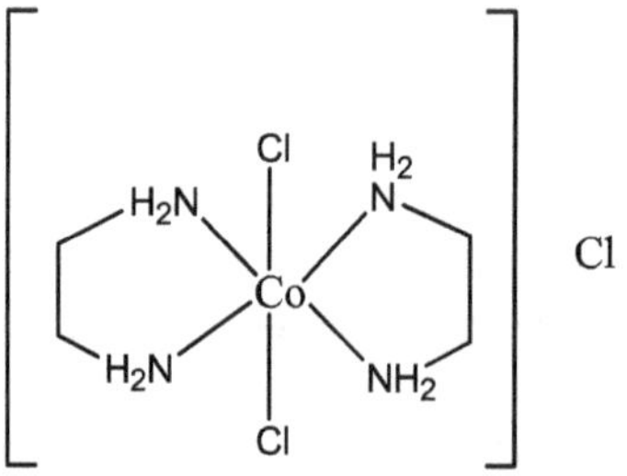

trans isomer

Fig. 3.8

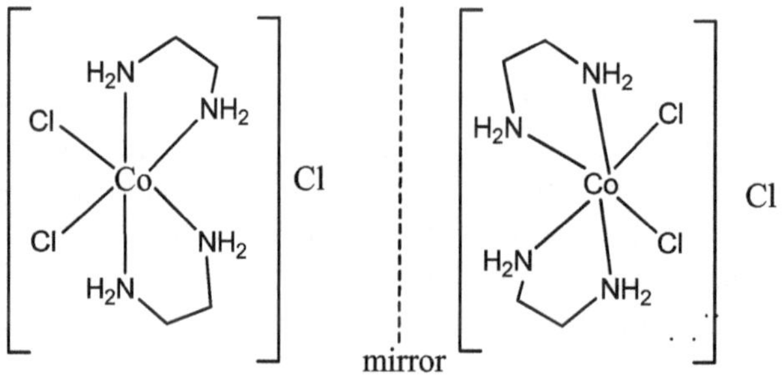

mirror

Cis-isomer has non-superimposable mirror images leads to enantiomers

Fig. 3.9

where en is ethylenediamine. Primarily, this complex has two isomers, namely *cis* and *trans* isomers. These two isomers arise from the position of two chloride ions in the complex. The *cis* isomer has two non-superimposable mirror images leading to optical isomers. Whereas, the *trans* isomer has a center of inversion and it does not form optical isomers. The *trans* isomer crystallises with one molecule of hydrochloric acid. Presence of hydrochloric acid in *trans* isomer is reflected in the elemental analysis and in the chloride estimation as well as in other spectroscopic data of the complex. Both the *cis* and *trans* form of the complex dichloro-*bis*(ethylenediamine)cobalt(III) chloride are diamagnetic.

The *trans* dichloro-*bis*(ethylenediamine)cobalt(III) chloride complex is prepared by the reaction of ethylenediamine with cobalt(II) chloride hexahydrate and hydrochloric acid in an oxidative condition maintained by purging oxygen.

$$CoCl_2.6H_2O \; + \; H_2N{\sim}NH_2 \; + \; HCl \; + \; O_2 \; \longrightarrow \; \left[\; \text{trans} \; \right] Cl.HCl \quad \quad .....(3.11)$$

Conversion of *trans*-dichloro-*bis*(ethylenediamine)cobalt(III) chloride to the *cis* form is brought about by evaporating a solution of the *trans*-dichloro-*bis*(ethylenediamine)cobalt(III) chloride to dryness on a steam bath. The unchanged *trans*-form can be removed by washing the mixture with cold water. Repeating the evaporation step completes the transformation. Heating should not be repeated more than two or three times, as some decomposition of the complex may take place on prolonged heating.

Synthesis of *cis* and *trans*-dichloro-bis(ethylenediamine)cobalt (III) chloride

To a solution of cobalt(II) chloride hexahydrate (1.60 g, 6.7 mmol) in water (5 ml) add a solution of ethylenediamine (6 g, 1.5 mmol) with stirring. Pass a vigorous stream of air through the solution for 10-12 hrs. Add concentrated hydrochloric acid (3.5 ml) to the reaction mixture and evaporate the solution on a steam bath until the volume reduces to approximately 10 ml. Cool the solution and allow it to stand overnight. Bright green square plates of the hydrochloride of the *trans*-dichloro-*bis*(ethylenediamine)cobalt(III) chloride will be formed. Filter the precipitate; wash it with ethylalcohol and diethylether. Dry the solid *trans*-dichloro-*bis*(ethylenediamine)cobalt(III) chloride at 110 °C. At this temperature the hydrogen chloride will be lost, and the crystals will transform to dull-green powder.

Oxalato complex

Oxalic acid is a dibasic acid. On deprotonation it forms dicarboxylate anion, which is a very good chelating ligand. The oxalate ligand with two negative charges, is used for the preparation of anionic complexes. The potassium *tris*-oxalatochromate(III) trihydrate complex is an example of anionic complex of oxalate as bidentate ligand. It is a complex of oxalic acid dianion with chromium at +3 oxidation state. The oxalate ligand and the complex has the structure as shown in Fig. 3.11.

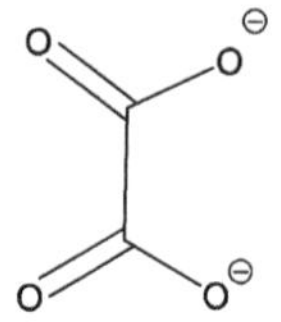

Oxalate dianion

Fig. 3.10

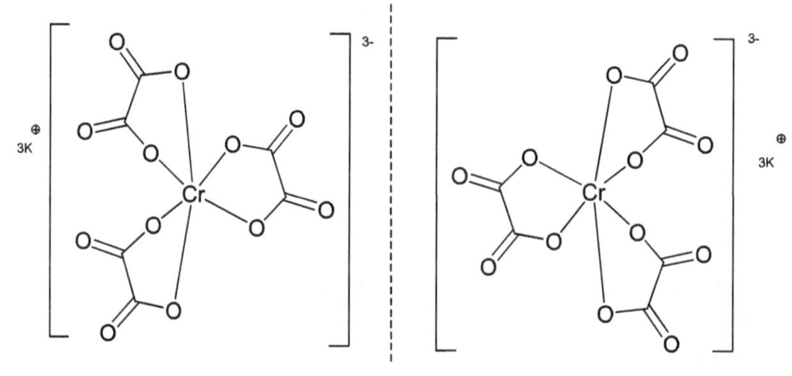

Fig. 3.11 *Tris*-oxalatochromate(III)

Potassium *tris*-oxalatochromate(III) trihydrate is synthesized from the reaction of potassium dichromate, oxalic acid and potassium oxalate. The following equation represents the synthesis:

$$K_2Cr_2O_7 + H_2C_2O_4 + K_2C_2O_4 \rightarrow K_3[Cr(C_2O_4)_3] + CO_2 + H_2O$$

$$.....(3.12)$$

It is well known fact that potassium dichromate oxidizes oxalic acid to carbon dioxide in acidic medium. In potassium dichromate the chromium is in +6 oxidation state. The chromium of dichromate gets reduced on reaction with oxalic acid to chromium with +3 oxidation state. The oxalate ligands bind to chromium ion at +3 oxidation state by forming a chelate complex anion. It is a six coordinated complex with octahedral environment. The complex is obtained from this reaction as a racemic mixture of two optical isomers, which can have non-superimposable mirror images. The complex occurs as deep green shiny crystals. The observed color of this compound strongly depends on the light, under which it is viewed. Under fluorescent light, the compound looks dark purple/black. When the compound is viewed under tungsten light, then it looks gray/black. In visible spectra the complex absorbs at 415 nm and at 570 nm. The visible spectra of the complex can be explained by using Orgel diagram. Electronic configuration of d-orbitals of chromium +3 in the complex is d^3 (t_{2g}^3 e_g^0). Free chromium ion has ground state term symbol ^4F. The Orgel diagram for d^3 system in an octahedral field is shown in Fig. 3.12.

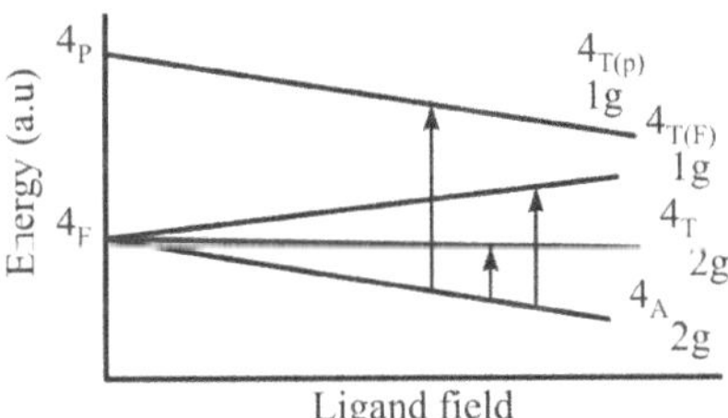

Fig. 3.12 Orgel diagram for d^3 system

Two absorptions observed at 570 nm is due to $^4A_{2g} \rightarrow {}^4T_{2g}$ and at 415 nm is due to $^4A_{2g} \rightarrow {}^4T_{1g}(F)$ transitions. The other electronic transition occurs at higher wavelength as a shoulder with low intensity. The low extinction co-efficients of these visible transitions are due to the fact that the d-d transitions are not allowed transition. It may be mentioned that the energy required for $^4A_{2g} \rightarrow {}^4T_{2g}$ corresponds to10Dq and it is about 17,000 cm^{-1}. The required Dq/B value for the ratio of the energy between the ratios of the first two electronic transitions namely $^4A_{2g} \rightarrow {}^4T_{2g}$ and $^4A_{2g} \rightarrow {}^4T_{1g}(F)$ transitions is 2.45. From this one may find out the Rachah parameter B.

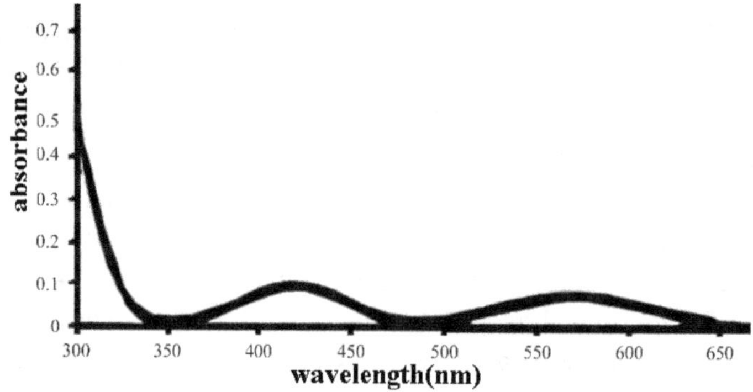

Fig. 3.13 UV-Visible spectra (MeOH) of potassium *tris*-oxalatochromate(III) trihydrate

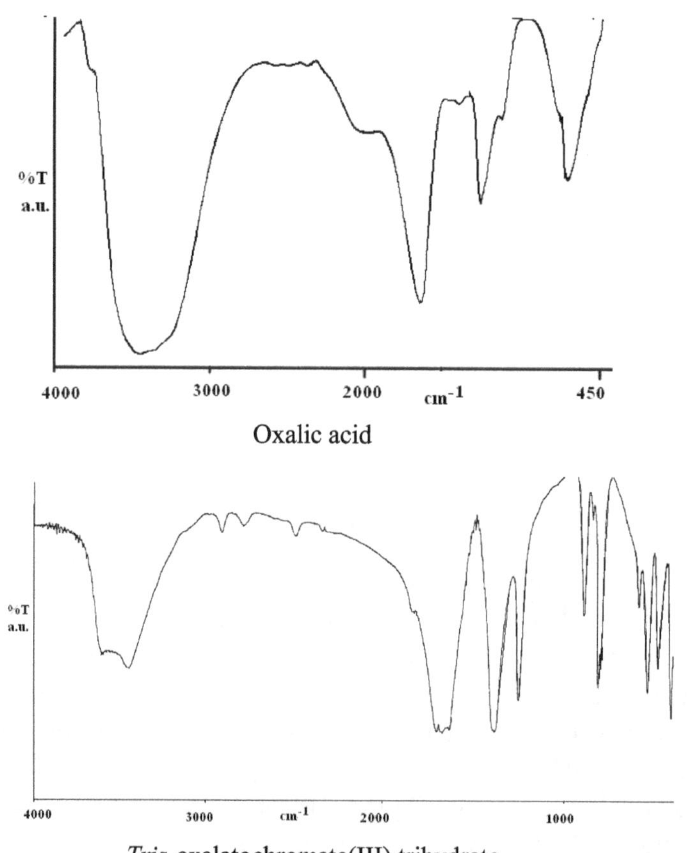

Oxalic acid

Tris-oxalatochromate(III) trihydrate

Fig. 3.14 FT-IR spectra (KBr, cm^{-1})

The calculated spin only magnetic moment value for the complex is 3.89 B.M. (based on 3 unpaired electrons) and the ground state of the chromium +3 is ^4A suggests that it should show spin only value for magnetic moment. The chromium complex has IR-absorptions at 1711 cm^{-1} for $v_{C=O\,(as)}$, 1680 cm^{-1} and 1634 cm^{-1} for $v_{C=O\,as}$, 1393 cm^{-1} for v_{C-O} and v_{C-C}, and at 1260 cm^{-1} for v_{symC-O} stretching frequencies respectively. For comparison, the IR spectra of the complex and that of the ligand namely oxalic acid are shown in Fig. 3.14.

Oxalic acid easily decomposes to carbondioxide under oxidative condition. Moreover, it is also thermally unstable and decomposes easily on heating. The metal carboxylate complexes are also thermally unstable. The potassium *tris*-oxalatochromate(III) trihydrate is no exception to this. The thermogravimetry is a technique in which weight of the compound vrs temperature is plotted. From thermogravimetric study the thermal stability, the way of thermal decomposition and kinetics of thermal reactions are stuied. The complex *tris*-oxalatochromate(III) trihydrate is a complex that systematically loses weight on heating due to its thermal instability. The percentage loss of weight of a weighed amount of the compound versus temperature is shown in figure 3.15.The compound thermally decomposes in two distinct steps. The initial weight losses corresponds to loss of the water of crystallization and in the next step the compound reacts with atmospheric oxygen and also loses carbon dioxide of the ligand to finally form a mixture of K_2O and Cr_2O_3.

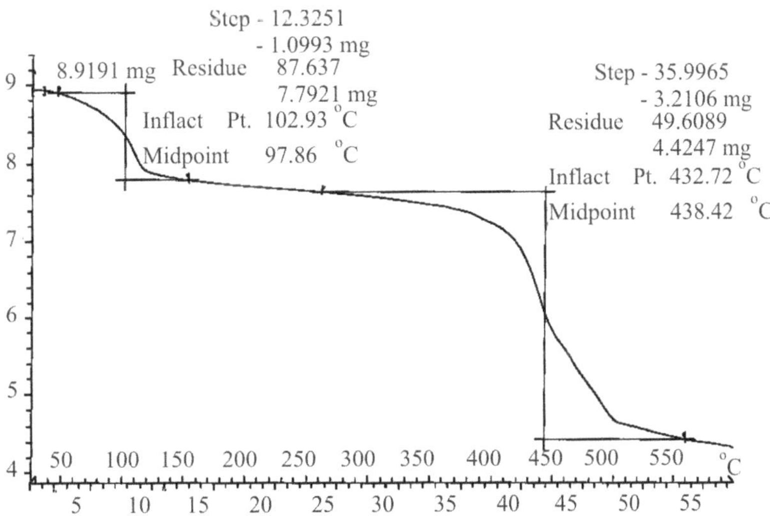

Fig. 3.15 Thermogram (10^{0}C / min) of potassium *tris*-oxalatochromate(III) trihydrate

The chemical composition or empirical formula for the complex can be arrived at easily by estimating chromium and carboxylate ligand in the complex. The amount of chromium in the complex can be determined by iodometric titration. For this purpose the chromium 3+ ion in the complex is oxidized to dichromate by boiling it with excess persulphate in the presence of silver nitrate catalyst.

$$2\ Cr^{3+} + 3\ S_2O_8^{2-} + 7H_2O \xrightarrow{\ \ AgNO_3\ \ } Cr_2O_7^{2-} + 6HSO_4^- + 8H^+$$

$$.....(3.13)$$

Boiling the solution for longer time destroys the excess persulphate remaining after the oxidation. The removal of excess persulphate from such solution before estimation of dichromate is required as persulphate being an oxidizing agent interferes in estimation of dichromate ion by iodometric titration. The dichromate content of the resulting solution is determined by titrating the solution with potassium iodide (excess amount). The liberated iodine in this reaction is titrated against standard sodium thiosulphate solution to find out the amount of dichoromate indirectly.

To estimate the amount of oxalate ligand in the complex, the complex is first decomposed in acidic condition by sulphuric acid as given in equation 3.14. The reaction leads to formation of oxalic acid and chromium sulphate. The chromium ions can interfere in the estimation of oxalic acid which is generally done by reaction with potassium permanganate under acidic condition.

$$K_3[Cr(C_2O_4)_3] + 6H^+ \xrightarrow{\qquad} Cr^{3+} + 3\ H_2C_2O_4 + 3K^+ + 3\ H_2O$$

$$.....(3.14)$$

The chromium +3 ions are precipitated as hydrated chromium oxide from the resultant solution by adding sodium hydroxide.

$$Cr^{3+} + OH^- \xrightarrow{\qquad} Cr_2O_3\ (H_2O)_n \downarrow$$

$$\text{Green ppt.} \qquad\qquad .....(3.15)$$

Excess sodium hydroxide should not be added while removing chromium from the solution as excess sodium hydroxide may dissolve again precipitated chromium trioxide. The green precipitate is to be filtered off and the filtrate is then titrated against potassium permanganate solution for the estimation of oxalate ions.

Synthesis and characterization of potassium tris-oxalatochromate (III) trihydrate

Make a solution of potassium oxalate monohydrate (2.3 g, 12.5 mmol) and oxalic acid dihydrate (5.5 g, 43.6 mmol) in water (110-120 ml). To this solution add solid potassium dichromate (1.9 g, 6.45 mmol) in small portions with constant stirring. Concentrate the solution nearly to dryness. On cooling deep green shiny crystals of potassium *tris*-oxalatochromate(III) trihydrate will be formed. Filter the solid and dry the solid by pressing between filter paper.

Estimation of chromium in the complex

Dissolve an accurately weighed potassium *tris*-oxalatochromate(III) trihydrate (in the range 0.10–0.15 g) in distilled water (100 ml). To this solution add sulphuric acid (5N, 5 ml), ammonium persulphate (~1 g) and silver nitrate solution (0.1M, 0.1 ml). Boil the solution for half an hour. The color of the solution turns to orange yellow at this stage. Cool the solution to room temperature. Estimate the chromium in the solution iodometrically by adding excess of potassium iodide, followed by titration of liberated iodine. Calculate the amount of chromium in the dissolved complex by assuming the complex in the solution to have transformed to dichromate.

Estimation of oxalate in the complex

Dissolve an accurately weighed potassium *tris*-oxalatochromate(III) trihydrate (in the range 0.10-0.15 g) in 20 ml distilled water. Add sulphuric acid (5N, 3 ml) to the solution and warm the solution. Carefully add dilute sodium hydroxide solution (50-70 ml). For this purpose, in the initial stage use a concentrated solution of sodium hydroxide and a dilute solution at the near neutral point. Continue addition of sodium hydroxide until no further precipitation of the hydrated chromium oxide is observed Addition of excess alkali should be avoided. Digest the solution by heating, filter the hot solution, and wash the residue thoroughly with water. Combine the filtrate and washings, acidify with sulphuric acid (5N, 15 ml), warm to 70 °C and then titrate the solution with standard potassium permanganate solution.

Glycine complexes

Amino acids have two binding sites one is carboxylic acid and other is amine. The amino acid ligands are capable of forming chelates. The amino acids can bind either through carboxylate anion or through neutral amino binding site. They can act as chelating ligand also. Amino acids as

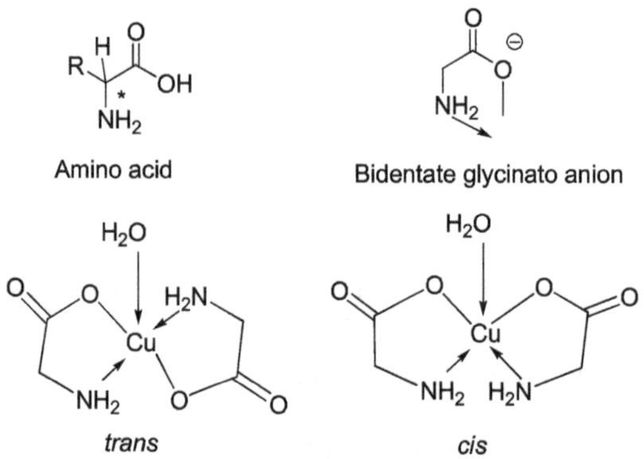

Fig. 3.16 Different binding modes of glycine

ligands have several advantages. They can form neutral complexes. Many naturally occurring amino acids are optically active. So, chiral metal complexes are formed with such amino acids. Amino acids can be functionalised at both the sides namely at the carboxylic acid site or at the amino group leading to different types of new ligands. They are the building blocks for proteins. Study of interactions of proteins with metal ions provides insight to their folding and leads to preparation of novel complexes to mimic biological activity. The copper complexes with amino acids are interesting. Glycine forms chelate complexes with copper(II) ion; two isomers of such copper(II) complexes are namely *cis* and *trans*-bisglycinatocopper(II) monohydrate. These two complexes are prepared by changing the reaction conditions. For example, the *trans*-bisglycinatocopper(II) monohydrate is prepared by solid-state reaction of copper(II) acetate monohydrate with glycine. Whereas, the solution phase reaction of copper(II) ions with glycine leads to the synthesis of *cis*-complex.

The IR spectra of glycine and *trans*-bisglycinatocopper(II) monohydrate are shown in Fig. 3.17. The complex *trans*-bisglycinatocopper(II) monohydrate has IR stretching absorptions due to v_{NH2} at 3315 cm^{-1} and 3263 cm^{-1}. The infra-red absorptions due to $v_{C=O}$ appears at 1602 cm^{-1}, deformation of NH_2 group δ_{NH2} frequency appears at 1619 cm^{-1}. The NH_2 region in the parent compound is broad. The broadening occurs due to merging of NH_2 absorption with OH absorption frequencies of carboxylic acid group. It also happens due to the contribution of IR absorptions of the NH_3^+ that arises from the zwitterionic form of the glycine.

Solid phase synthesis of *trans-bis*glycinato copper(II) monohydrate

Mix accurately weighed amount of copper (II) acetate monohydrate (2 g, 10 mmol) and glycine (2 g, 26.6 mmol) in an agate mortar at room temperature. Grind the mixture thoroughly in the mortar. Transfer the paste to a watch glass and leave it for 2 hrs. The color of the mixture will change from green to pale blue. In 2 hrs, all the reactants will be exhausted. Wash the *trans-bis*glycinato copper(II) monohydrate with alcohol followed by ether and dry it in vacuum.

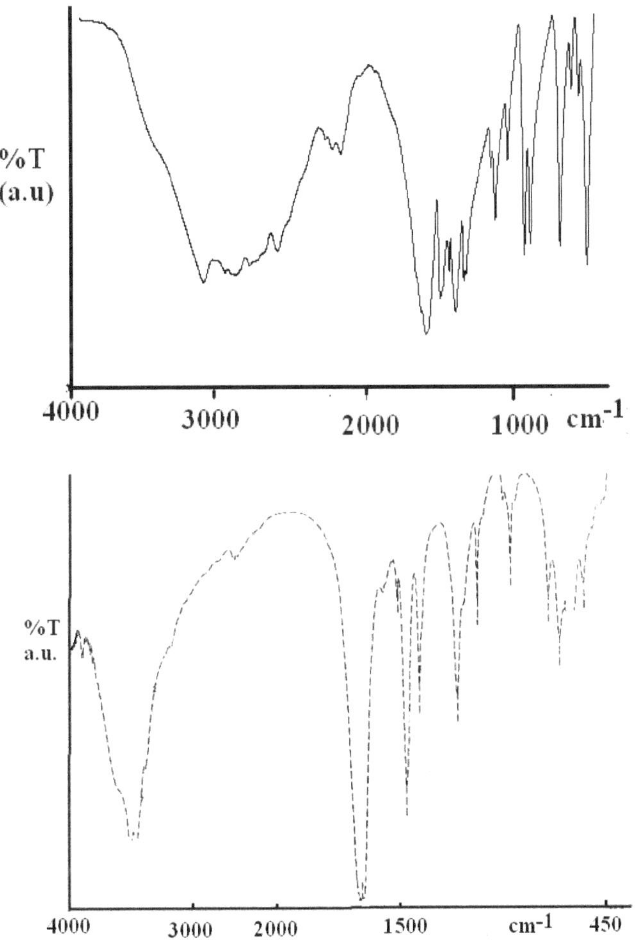

Fig. 3.17 FT-IR spectra (KBr, cm^{-1}) of glycine (top) *and trans-bis*-glycinato copper(II) (bottom)

Estimation of copper in *trans- bis*glycinato copper(II) monohydrate

Take an accurately weighed (0.15 g) amount of *trans*-bis glycinato copper(II) monohydrate in sulphuric acid (1M). Heat the solution so that the complex decomposes to form copper(II) sulphate. Add excess of potassium iodide to the copper(II) sulphate solution thus formed. Carry out volumetric titration with standard sodium thiosulphate solution to estimate the iodine liberated by reaction of copper(II) with potassium iodide and calculate the amount of copper in the dissolved sample.

Phosphine complexes

The triphenylphosphine is one of the most important phosphine ligand which binds to metal ions in monodentate fashion. It forms complexes in different oxidation states of metal ions. The triphenylphosphine complex of rhodium at +1 oxidation state with a molecular formula chloro *tris*-triphenylphosphine rhodium(I) is a catalyst for various organic reactions. The triphenylphosphine plays several roles in catalytic role of this complex (i) triphenylphosphine complex makes the complex more soluble in organic solvent, so the complex becomes suitable for homogeneous catalysis of organic reactions. (ii) It stabilizes low oxidation states of the metal ion. (iii) Due to its bulky nature, it can control the reactivity of substrate through steric effects. (iv) Its relative lability allows exchange of ligands. The $Rh(PPh_3)_3Cl$ has four coordination number and it has a tetrahedral structure. The compound is prepared by reacting rhodium(III) chloride polyhydrate with ethanol

$$RhCl_3. xH_2O + CH_3CH_2OH + PPh_3 \rightarrow RhCl(PPh_3)_3 + CH_3CHO +$$

$$HCl + H_2O \qquad\qquad\qquad .....(3.16)$$

in the presence of triphenylphosphine. In this reaction ethanol serves as a reducing agent and it gets oxidised to acetaldehyde. The complex is a 16-electron complex.

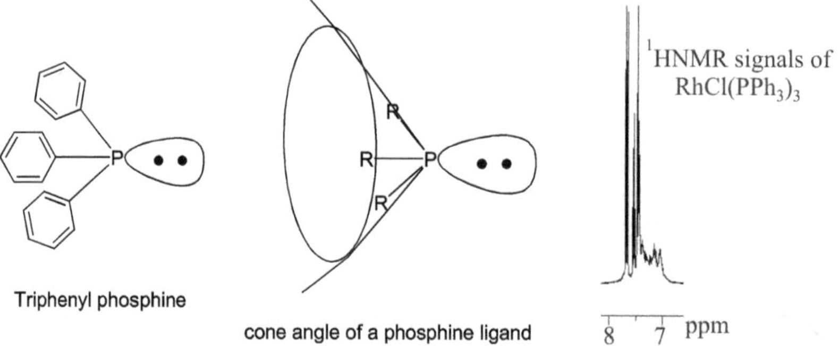

Fig. 3.18 Left: Triphenylphosphine ligand and its cone angle; Right: ^{1}HNMR (in CDCl$_3$) of RhCl(PPh$_3$)$_3$ (aromatic region only)

The chloro *tris*-triphenylphosphine rhodium(I) has characteristic proton nuclear magnetic resonance (^{1}HNMR) signals in the aromatic region that arises from the aromatic protons of the triphenylphosphine ligands as illustrated in Fig. 3.18. The triphenylphosphine ligands of RhCl(PPh$_3$)$_3$ are labile and these ligands dissociates easily to give a red-colored binuclear complex Rh$_2$Cl$_2$(PPh$_3$)$_4$. The Rh$_2$Cl$_2$(PPh$_3$)$_4$ has very poor solubility in benzene, so the formation of it can be easily identify on the basis of solubility. The RhCl(PPh$_3$)$_3$ is highly soluble in solvents like benzene, toluene, dichloromethane, chloroform etc. The RhCl(PPh$_3$)$_3$ reacts with aldehydes to eliminate carbon monoxide and which is trapped in the form of a rhodium complex having composition RhCl(CO)(PPh$_3$)$_2$. The reaction is illustrated in equation 3.17.

$$RhCl(PPh_3)_3 + RCHO \rightarrow RhCl(CO)(PPh_3)_2 + RH + PPh_3$$

$$.....(3.17)$$

Chloro *tris*-triphenylphosphine rhodium(I) is known as Wilkinson's catalyst and for its discovery and its role in catalysts a Nobel prize was awarded. It is a versatile catalyst for hydrogenation, isomerisation of olefins, hydroboration, hydrosilylation reaction etc.

Square planar complexes

The complex Ni(PPh$_3$)$_2$Cl$_2$ is an important complex for its catalytic reactivity for C-C bond formation. It has four coordination number and it has a four coordinated geometry. It belongs to ML$_2$X$_2$ type of four coordinated complex (L – mondentate ligand, X = halide); such four coordinated complexes can adopt either tetrahedral or square planar geometry. A square planar complex can have cis and trans form. Some of the structures that are generated from four coordinated ML$_2$X$_2$ type of complexes are shown in Fig. 3.19.

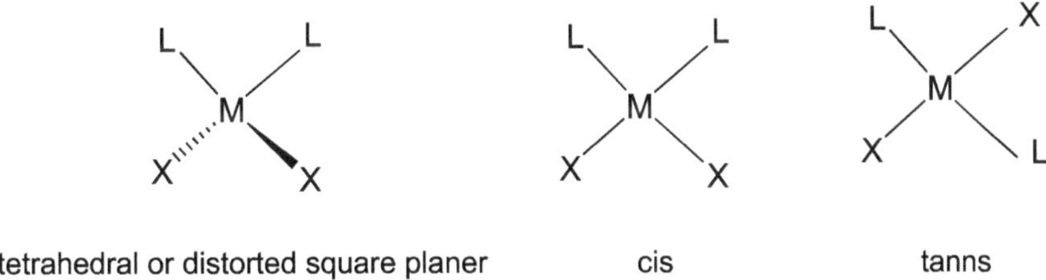

Fig. 3.19 Isomers of 4-coordination number complexes

The complex Ni(PPh$_3$)$_2$Cl$_2$ has a distorted tetrahedron structure. Generally nickel(II) with d^8 electronic configuration prefers square planar geometry, but due to the two bulky triphenylphosphine ligands and two chloride ligands, this particular complex takes the shape of distorted tetrahedron. The complex is synthesized by the following reaction:

$$NiCl_2\ 6H_2O + PPh_3 \xrightarrow{\text{Acetic acid}} [Ni\ (PPh_3)_2Cl_2] \qquad .....(3.18)$$

Similar complexes of $Ni(bpy)Cl_2$, $Ni(py)_2Cl_2$ (where bpy is 2,2′-bipyridine, py is pyridine) are also prepared in a similar manner. However, these complexes are prepared by reacting anhydrous nickel(II) chloride with the corresponding ligand in isopropanol as solvent.

Preparation of *bis*-chloro *bis*-triphenylphosphine nickel(II)

To a solution of nickel(II) chloride hexahydrate (1.19 g, 5 mmol) in water (2 ml), add glacial acetic acid (50 ml). To this solution add a solution of triphenylphosphine (2.62 g, 10 mmol) dissolved in acetic acid (15 ml). Olive green microcrystalline precipitate will be formed on standing. Keep the precipitate in contact with the mother liquor for 24 hrs. Dark green crystals will appear. Separate the crystalline solid by filtration and collect the solid, dry them by pressing between filter papers.

Estimation of Nickel

The amount of nickel in the complex *bis*-chloro *bis*-triphenylphosphine nickel(II) is estimated by gravimetric method. It is done by reacting a known amount (about 20 mg) of *bis*-chloro *bis*-triphenylphosphine nickel(II) with concentratted nitric acid. Followed by making the solution basic (pH 9) and adding dimethylglyoxime (dmg) to the solution. The *bis*-dimethylglyoximato nickel(II) complex is formed (Fig. 3.20) and gets precipitated as pink solid. Filter the solid to a pre-weighed sintered crucible. Dry the precipitate at 110 °C and determine the weight of *bis*-dimethylglyoximato nickel(II) complex. From this quantity determine the amount of nickel present in the complex.

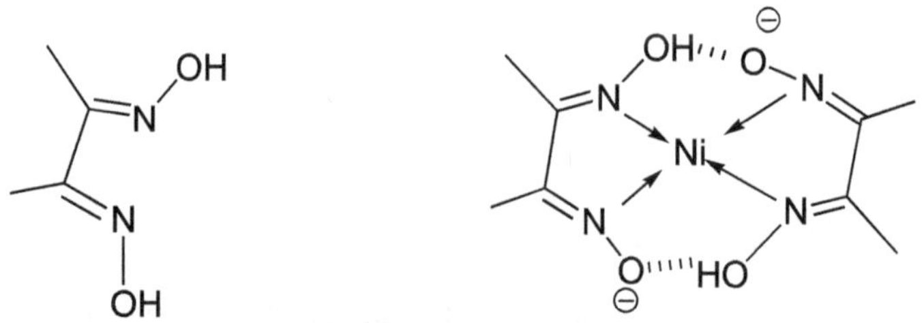

Dimethylglyoxime   ***Bis*-dimethylglyoximato nickel(II)**

Fig. 3.20

Cis and trans isomers in platinum complexes

Platinum complexes in +2 oxidation state with four coordination number generally adopt square planar structures. Some of the four coordinated platinum complexes find application in cancer therapy. Many platinum complexes are used as catalyst for organic reactions. The platinum complex with ethylene as ligand has led to understanding of olefin binding to metal ions and opened the field of orgnanometallic chemistry with olefin compounds.

One of the most well known and important platinum complex is *bis*-ammine di-chloro platinum(II). It is commonly known as cisplatin. It is used for cancer treatment. It is prepared by the following methods:

(a) by adding ammonia to ammonium tetra-chloro platinum(II);

(b) by heating tetra-ammine platinum(II) chloride at 250°C; or

(c) by reacting ammonium carbonate with tetra-chloro platinic(II) acid

The platinum(IV) salts are commonly available and they are first converted to platinum(II) for performing chemistry at +2 oxidation state. While preparing cisplatin from potassium hexa-chloro platinate(IV) the first step is the reduction of platinum(IV) to platinum(II) by hydrazine hydrochloride.

$$K_2[PtCl_6] \; + \; N_2H_4.HCl \; \longrightarrow \; K_2[PtCl_4] \; + \; N_2 \; + \; HCl$$

$$.....(3.19)$$

In the second step the tetra-chloro platinum(II) is treated with ammonia and ammonium chloride to obtain exclusively cisplatin. This happens due the effect called trans-effect.

$$K_2[PtCl_4] \; + \; NH_3 \; \xrightarrow{NH_4Cl} \; cis\text{-}[Pt(NH_3)_2Cl_2] \qquad .....(3.20)$$

The transplatin is prepared by the reaction of ammonia with potassium tetra-chloro platinate(II).

$$K_2[PtCl_4] \; + \; NH_3 \; \longrightarrow \; [Pt(NH_3)_4](OH)_2 \; + \; HCl \longrightarrow \; trans\text{-}[Pt(NH_3)_2Cl_2]$$

$$.....(3.21)$$

The *cis* and *trans* compounds are distinguished by chemical reactivity of these complexes. A hot aqueous solution of the *cis* complex reacts with aqueous thiourea to give a deep yellow solution. On standing, this complex forms yellow *tetrakis*-thiourea platinum(II) chloride. The needles like crystals of *tetrakis*-thiourea platinum(II) chloride can be obtained from the yellow solution. The *trans* compound from similar reaction gives a colorless solution of *trans-bis*thiourea di-ammine platinum(II). On crystallization *trans-bis*-thiourea di-ammine platinum(II) is obtained as snow-white needles.

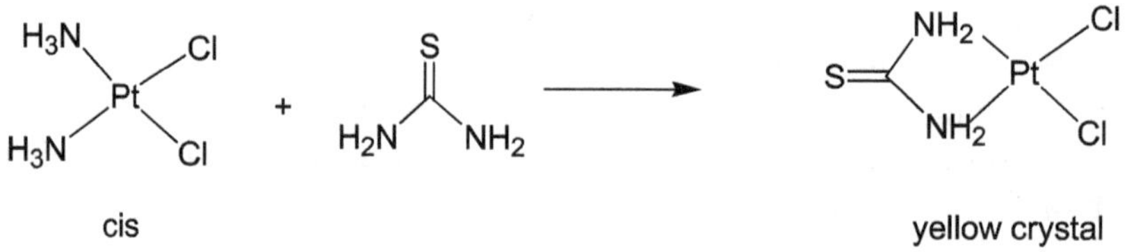

cis   +   →   yellow crystal

$$.....(3.22)$$

trans   +   →   white crystal

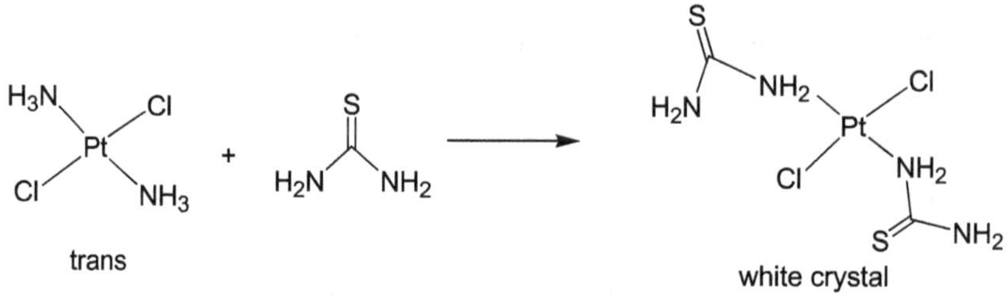

$$.....(3.23)$$

The *trans* effect is the ability to direct an incoming ligand to its *trans* position while ligand substitution reaction taking place.

For example, the reaction for the substitution of the chloride ligands with ammonia in tetra-chloro platinate dianion occurs due to *trans*-effect. The substitution of the first chloride ligand in this reaction is non-specific. But, the second substitution reaction in the complex occurs at *cis* position with respect to the first substituted ligand, which in this case is an ammonia ligand. This happens due to the greater *trans*-effect of the chloride ligand relative to the ammonia ligand. Such *trans* directing effect of a ligand are arranged in order of reactivity and the series called the *trans*-effect series. The *trans*-ability is shown in following order:

CN^-, CO, NO, C_2H_6 > PR_3, H^- > CH_3^-, $C_6H_5^-$, $SC(NH_2)_2$, SR_2 > SO_3H^- > NO_2^-, I^-, SCN > Br^- > Cl^- > pyridine > RNH_2, NH_3 > OH^- > H_2O

The magnitude of this effect is determined by the δ-donor or π-acceptor ability of the ligands *trans* to the leaving ligand. The sigma effect arises due to the possible formation of a trigonal bipyramidal transition state. For the sigma-trans-effect, a polarisable strong trans-sigma-donor ligand need to have its electron density distributed more towards the leaving ligand. The *trans* directing ligand should have greater overlap with the empty p-type of σ-orbital. Such overlap lowers the energy of the transition state. Few sigma-trans-directing ligands are: H^- > PR_3 > $- SCN^-$ > I^- > CH_3^- > CO > CN^- > Br^- > Cl^- > NH_3 > OH^-

For the π-trans-effect, a trans-π-acceptor ligand, should form π-bonds with the metal and withdraw electron density away from the leaving group making it more labile. As a consequence of such effect, the electron density around the leaving group will attract incoming nucleophilic ligands, and it also stabilizes the transition state with trigonal bipyramidal geometry. Some π-*trans*-directing ligands are:

$$CH_2 = CH_2 > CO > CN^- > -NO_2 > -SCN^- > I^- > Br^- > Cl^- > NH_3 > OH^-$$

The σ- and π- effects contribute to lowering the energy of the transition state which favors the preferential substitution of a ligand at a *trans* position.

Preparation of potassium tetra-chloro platinate(II)

Make a suspension of (0.97 g, 0.002 mol) of potassium hexa-chloro platinate(IV) in 10 ml water, in a 100 ml beaker. To this suspension add (0.10 g, 0.01 mol) hydrazine hydrochloride. In this step excess hydrazine hydrochloride should not be added, excess addition of hydrazine hydrochloride can cause reduction of platinum(IV) to metallic platinum. Stir the mixture so that the temperature is raised to about 60 °C for 5-10 mins. Leave the reaction mixture over a water bath at this temperature for about 2 hrs, until only a small amount of yellow potassium hexa-chloro platinate(IV) remains in the deep red solution. Now, raise the temperature to 90 °C to complete the reaction. Cool the mixture in an ice bath and filter to remove unreacted potassium hexa-chloro platinate(IV). Wash the residue with several portions (2 ml) of ice-water until the washings are colorless. The washings, combined with the deep red filtrate, contain pure potassium tetra-chloro platinate(II) in dilute hydrochloric acid.

Preparation of *cis*-[Pt(NH₃)₂Cl₂]

Take half the portion of the potassium tetra-chloro platinate(II) solution obtained in the earlier experiment and add ammonium chloride (0.3 g) in a 50 ml beaker. Stir it to dissolve the ammonium chloride and add 1.5 ml of aqueous ammonia to this solution to make the solution basic. Excessive use of ammonia will decrease the yield of tetra-ammine-platinum(II) chloride. Keep the solution in a refrigerator for 24 hrs. A greenish yellow solid precipitate will appear. During this period, the supernatant liquid changes from deep red to light yellow. The precipitate, consists of the *cis* isomer and a small amount of tetra-ammine platinum(II) tetra-chloro platinate(II). Filter the precipitate and wash it with portions of ice water to make free from soluble salts of platinum(II). Transfer the precipitate to a beaker containing 0.01N hydrochloric acid

(15 ml). The cisplatin gets dissolved and remains in solution/filtrate; discard the residue. On concentration of the filtrate cisplatin is obtained.

Preparation of *trans*-[Pt(NH₃)₂Cl₂]

To prepare transplatin add excess ammonia to an acid solution of potassium tetra-chloro platinate(II). Stir the solution for few hrs to obtain tetra-ammine platinum(II) chloride. Heat the solution in the presence of excess hydrochloric acid to obtain transplatin.

Metal complexes at low oxidation states

Triphenylphosphine has the extra ability to stabilize low oxidation states of a metal ion through back bonding. In this case the back bonding to metal occurs from filled d-orbitals of a metal ion to the vacant d-orbitals of phosphorous (Fig. 3.21). The phosphine complexes are used for various catalytic reactions as they can stabilize low oxidation state of metal. The metal complexes at low oxidation state has tendency to form metal carbon bonds.

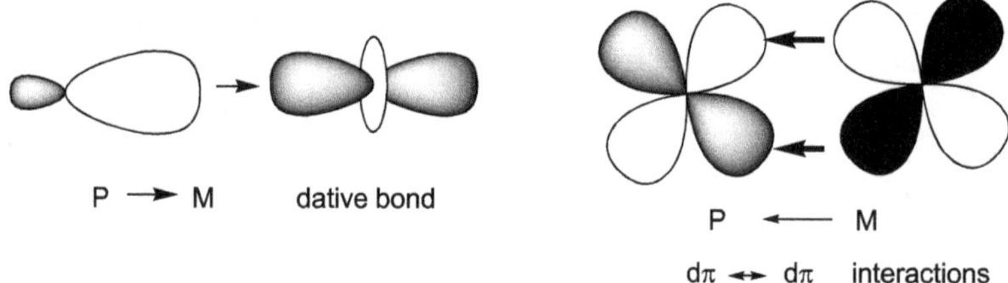

Fig. 3.21 Synergic dative and back bond formation

The copper(I) complex *tris*-triphenylphosphine copper(I) nitrate with triphenylposphine as a ligand is formed by reaction of copper(II) nitrate with triphenylphosphine in alcohol. In this reaction triphenylphosphine gets oxidized and copper(II) ion gets reduced to copper(I). The complex is colorless and it is diamagnetic complex. The complex has copper(I) with a d^{10}-electronic configuration. The complex has its characteristic IR-absorptions from triphenylphosphine ligand. The complex has a monodentate nitrate group as evident from the IR signals at 1434 cm^{-1}, 1303 cm^{-1}, 1091 cm^{-1} that are characteristic of monodentate nitrate group. Moreover, the complex is non-ionic in nature in acetonitrile solvent showing that it is a neutral complex. The nitrate binds to the copper ion in a monodentate fashion. The amount of copper in the complex is estimated by converting copper(I) to copper(II), followed by estimation of copper (II) by iodometric method.

Synthesis and characterization of *tris*-triphenylphosphine copper(I) nitrate

Dissolve a solution of copper(II) nitrate trihydrate (0.365 g, 1.5 mmol) in minimum amount of ethanol. Make another solution by dissolving triphenylphosphine (1.965 g, 7.5 mmol) in 30 ml ethanol under warm condition. Mix the two solutions, reflux the solution for half an hr and cool the reaction mixture. On cooling a white precipitate will appear. Filter the white crystalline solid and wash with ethanol. For obtaining good crystalline complex it is necessary recrystallize the white solid from methanol.

Estimation of copper

Decompose an accurately weighed (0.10-0.15 g) copper complex *tris*-triphenylphosphine copper(I) nitrate, by heating with concentrated nitric acid (1 ml). To the resulting oily mass, add water (30 ml) and filter. Treat the filtrate with sodium hydroxide (2M, 50 ml) to precipitate copper hydroxide. Filter the solution and reject the mother liquor. Wash copper hydroxide precipitate with water and dissolve it in dilute sulphuric acid (10 ml, 1M). The resulting solution contains copper(II) sulphate, estimate the copper present in the solution by treating with excess potassium iodide. The liberated iodine is titrated with sodium thiosulphate to find out the amount of copper that liberated the iodine.

Pyridinium tetrachlorocuprate (II)

The pyridinium tetra-chloro cuprate is an example of four coordinated copper(II) complex. The complex, as such tetra-chloro cuprate complexes are important from structural point of view; they show thermochromic properties i.e. they change color with temperature. The complex pyridinium tetra-chloro cuprate is prepared by dissolving copper(II) chloride in methanol, followed by acidification with concentrated hydrochloric acid. This makes the solution brown due to formation of $H_2[CuCl_4]$. Addition of pyridine to this solution, the desired complex is precipitated. The complex has a spin only magnetic moment value 1.78 BM. This magnetic moment value arises due to single unpaired electron of d^9 system. The UV-visible absorption of this complex is highly solvent dependent. In aprotic solvent like acetonitrile it absorbs at 474 nm whereas it absorbs at 437 nm in methanol. In solid state the complex is yellow at room temperature but it is green at liquid nitrogen temperature. The color change of the complex with temperature is attributed to the fact that at room temperature the complex has a distorted tetrahedral geometry, which flattens to square pyramidial geometry at low temperature. Such geometrical changes are important in making sensors. For example, structural changes in nickel complexes that have fluorescence active ligands are capable of measuring temperature with accuracy upto third digit. This is done by monitoring the changes in fluorescence emission of the complex with temperature, as fluorescence is very sensitive technique a small change in temperature can be detected. Such small changes in temperature are not possible to detect by a conventional mercury thermometer. The pyridinium tetra-chloro cuprate(II) complex also shows interesting thermo chemical changes on heating to higher temperature (scheme 3.1).

$$(PyH)_2[CuCl_4] \quad \rightarrow \quad Cu(py)_2Cl_2 \quad \rightarrow \quad CuCl_2$$

Scheme 3.1 Thermal degradation of pyridinium tetrachlorocuprate(II)

On heating pyridinium tetra-chloro cuprate(II) complex first loses two molecules of hydrochloric acid. In a second stage it loses two pyridine molecules as illustrated in the scheme 3.1.

The complex has characteristic IR absorption for $v_{C=C}$ and $v_{C=N}$ starching absorptions at 1600 cm^{-1} and 1525 cm^{-1} respectively. The N–H vibration appears at 2850 cm^{-1}, whereas absorption for N–H vibration of the hydrochloric acid salt of pyridine namely pyridinum hydrochloride appears at 3200 cm^{-1}.

Carboxylate complexes

Carboxylic acid is an important ligand and it forms large number of metal complexes. The metal complexes of anion of carboxylic acids are known as metal carboxylates. Metal carboxylates have interesting structural chemistry. They find major applications in biology and in material science. The carboxylate complexes are prepared by reacting a carboxylic acid with a metal hydroxide or with metal carbonates. For example, evaporation of a solution/ suspension of copper(II) carbonate in acetic acid gives the copper(II) acetate monohydrate.

$$CuCO_3 \cdot xH_2O + CH_3COOH \rightarrow [Cu_2(CH_3COO)_4] \cdot 2H_2O + CO_2$$

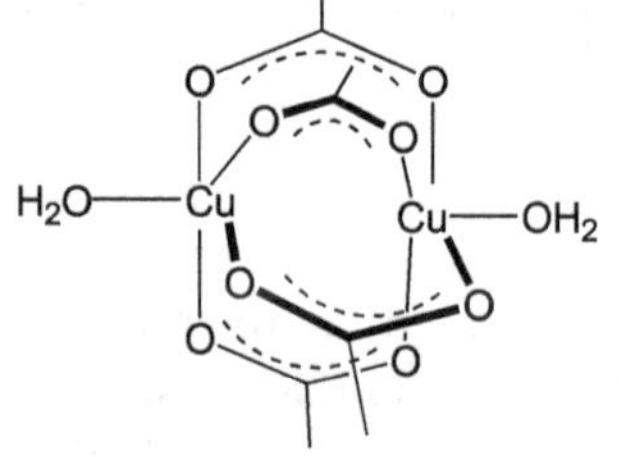

Dinuclear structure copper(II) acetate monhydrate

Fig. 3.22 Copper(II) acetate dimer

Besides the monodentate, chelating bidentate binding modes of carboxylates; the carboxylate ligands have different types of bridging binding modes to hold multiple number of metal ions, some of such simple modes are shown in Fig. 3.23. These binding modes make their coordination chemistry diverse. They find useful in construction polymeric and network structures. Such three dimensional network structures are of great importance as porous materials.

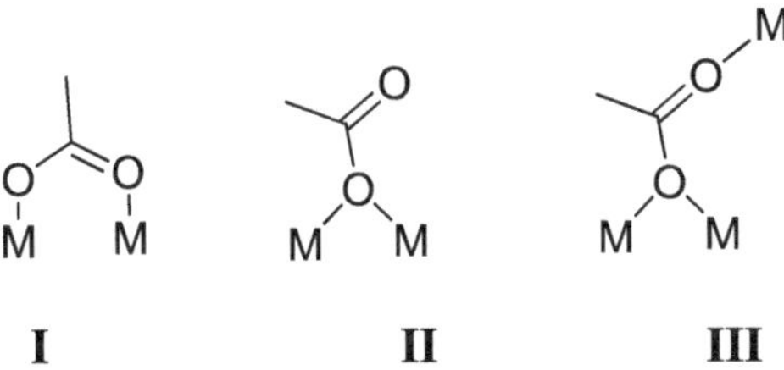

Fig. 3.23 Different bridging modes of carboxylate group

The metal carboxylates can adopt either mono-nuclear, dinuclear, or polynuclear carboxylate complexes depending on reaction conditions. Some binuclear metal carboxylate complexes are prepared by ligand replacement reactions.

Chromium(II) acetate, exists as dimer having a composition $[Cr_2(CH_3CO_2)_4(H_2O)_2]$. It is a diamagnetic compound. This is an example of complex that shows quadruple bond. $[Cr_2(CH_3CO_2)_4(H_2O)_2]$ is non-ionic, with poor solubility protic solvents. It has a dimeric structure in which each half of the dimeric structure consists of one chromium ion coordinated to four oxygen atoms (one each from acetate group) to form a square, along with one water molecule in axial position. The chromium atoms having similar environment are bonded through chromium-chromium bond to make an octahedral geometry. The molecule has D_{4h} symmetry. The quadruple bond between the two chromium atoms arises from the overlap of four d-orbitals from the chromium ions. The d_z^2 orbital overlap to form a sigma bond, whereas the d_{xz} and d_{yz} orbitals overlap to give two π-bonds and the d_{xy} bond forms the delta bond. Intermolecular distance between the two chromium atoms in this complex is 2.36Å.

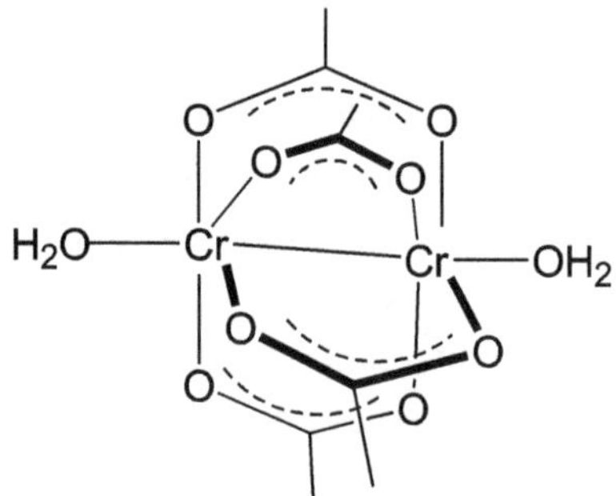

Fig. 3.24 Structure of chromium(II) acetate hydrate

The chromium(II) acetate is prepared from aqueous solutions of chromium (III) salts. For this purpose chromium (III) ions are reduced to the chromous state by zinc dust. The reduction leads to blue solution which on reaction with sodium acetate gives the precipitate of chromium(II) acetate as a bright red powder. The overall reaction is as follows:

$$Cr^{3+} + Zn \rightarrow Cr^{2+} + Zn^{2+} \qquad .....(3.23)$$

$$2\, Cr^{2+} + 4\, OAc^- + 2\, H_2O \rightarrow [Cr_2(OAc)_4\,(H_2O)_2] \quad .....(3.24)$$

Anhydrous $[Cr_2(OAc)_4]$ is synthesised from chromocene by reacting with acetic acid.

Chromocene $+\ CH_3COOH \longrightarrow$ anhydrous chromous acetate

$$.....(3.25)$$

$[Cr_2(OAc)_4(H_2O)_2]$ is occasionally used to dehalogenate organic compounds such as α-bromoketones. The chromous acetate is a good reducing agent and it gets oxidised by single electron transfer, so it is used to reduce oxygen in air.

Magnetic moment measurement

Magnetic moment of an inorganic complex can simply be measured on a magnetic susceptibility balance. Insert an empty pre-weighed sample tube to the tube guide of the instrument, measure the magnetic deflection caused by the tube (R_0). Take a weighed amount of the finely ground solid sample in a sample tube. Insert the sample tube with the sample to the tube guide and record magnetic deflection caused by the tube (R). Calculate the mass susceptibility, χ_g using equation given below:

$$\chi_g = \frac{C_{bal}\ 1\ (R - R_0)}{10^9\ m} \qquad \dots(3.26)$$

where,

    1 = Sample length (cm)

    m = Sample mass (g)

    R = Reading for tube plus sample

    R_0 = Empty tube reading

    C_{bal} = Balance calibration constant

The molar susceptibility,

$$\chi_m = \chi_g\ M$$

where, M = molecular formula weight of the substance

For compounds containing a paramagnetic ion, χ_M are less than susceptibility / g atom of the paramagnetic ion, χ_A, because of the diamagnetic contribution of the other groups or ligand present. Since magnetic moments is additive, χ_A can be obtained from χ_M by adding appropriate corrections.

$$\chi_A = \chi_M - \Sigma \chi_l \qquad \dots(3.27)$$

where, $\Sigma \chi_l$ = diamagnetic correction.

The effective magnetic moment (μ_M) is calculated from the equation below

$$\mu = 2.84\ \chi_M\ T \qquad \dots(3.28)$$

Cobalt(II) benzoate complexes

The reactions of cobalt(II) chloride or cobalt(II) acetate with sodium benzoate in the presence of pyridine can lead to different types of benzoate complexes depending on the reaction conditions. For example the reaction of benzoic acid with cobalt(II) chloride followed by reaction with pyridine gives a dinuclear mixed aqua and benzoate bridged cobalt(II) complex that has a benzoic acid as solvent of crystallization (Fig 3.25).

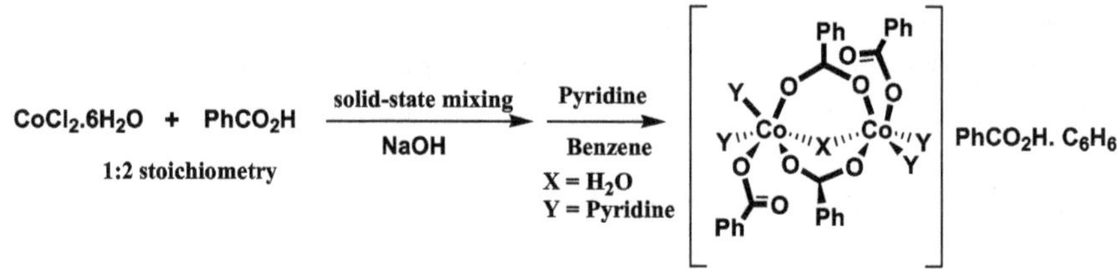

$$CoCl_2.6H_2O \ + \ PhCO_2H \xrightarrow[\text{NaOH}]{\text{solid-state mixing}} \xrightarrow[\substack{\text{Benzene}\\ X = H_2O\\ Y = \text{Pyridine}}]{\text{Pyridine}} \left[\text{(dinuclear Co complex)} \right] PhCO_2H.\ C_6H_6$$

.....(3.29)

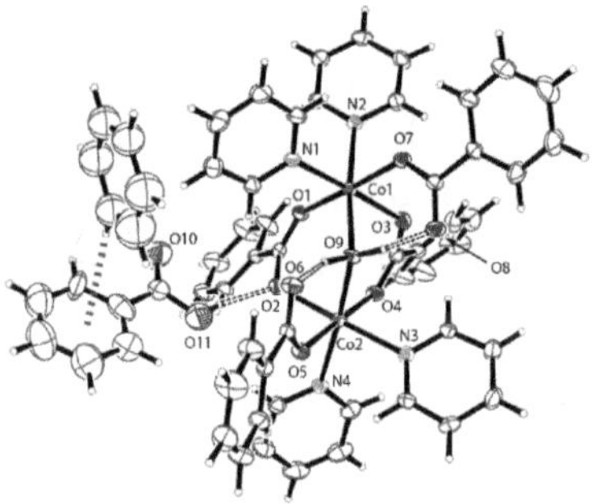

Fig. 3.25 A dinuclear cobalt(II) complex with benzoic acid as solvent of crystallization

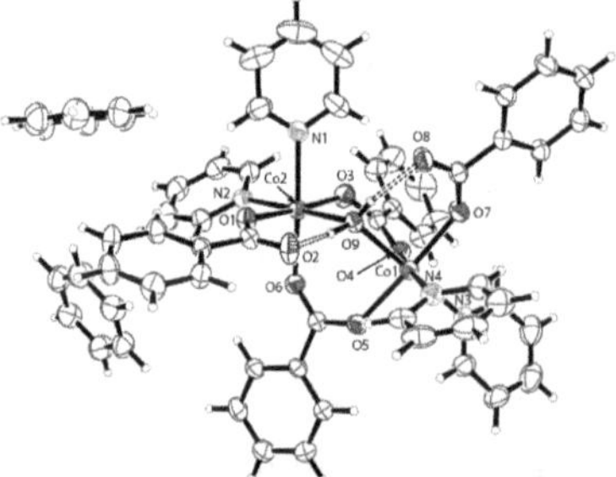

Fig. 3.26 A dinuclear cobalt(II) complex with benzene as solvent of crystallization

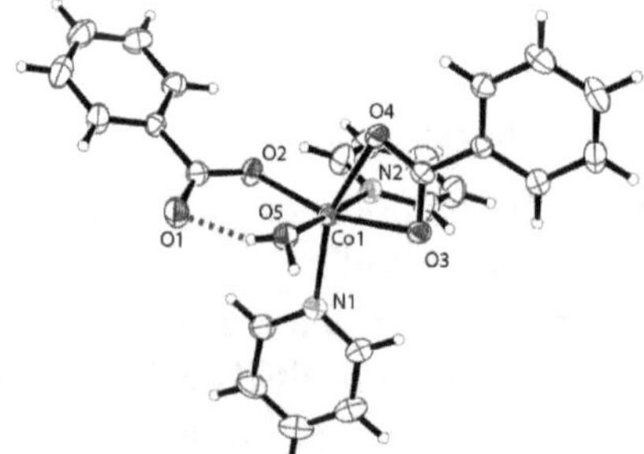

Fig. 3.27 A mononuclear aqua-*bis*-pyridine di-benzoatozinc(II)

Similar reaction of sodium benzoate with cobalt(II) chloride followed by reaction with pyridine in benzene gives a similar aqua and benzoate bridged dinuclear cobalt(II) complex that has benzene as a solvent of crystallization (Fig. 3.26 and equation 3.30).

$$CoCl_2 \cdot 6H_2O + PhCO_2Na \xrightarrow[\text{1:2 stoichiometry}]{\text{solid-state mixing}} \xrightarrow[\substack{\text{Benzene} \\ X = H_2O \\ Y = Pyridine}]{\text{Pyridine}}$$

.....(3.30)

Mononuclear cobalt(II) benzoate complex aqua-*bis*-pyridine-di-benzoato zinc(II) complex is prepared by reacting benzoic acid with cobalt(II) acetate and pyridine in methanol (Fig. 3.27 and equation 3.31).

$$Co(OAc)_2 \cdot 6H_2O + PhCO_2H \xrightarrow[\substack{\text{Pyridine} \\ X = H_2O \\ Y = Pyridine}]{\text{MeOH}}$$

.....(3.31)

Synthesis of [Co$_2$(μ-H$_2$O)(μ-OBz)$_2$(OBz)$_2$(Py)$_4$].(C$_6$H$_6$)(BzOH)

Finely grind in mortar a mixture of benzoic acid (0.24 g, 2 mmol), sodium hydroxide (0.08 g, 2 mmol) and cobalt(II) chloride hexahydrate (0.238 g, 1 mmol) and heat the mixture at 100 °C for 15 min, the pink color will turn blue. Cool the solid mixture to room temperature and transfer it into a round bottom flask. Add benzene (20 ml) to it and stir the heterogeneous mixture at room temp for 5 min followed by the addition of pyridine (0.158 g, 2 mmol). The blue supernatant liquid will turn pink. Filter off the residue and leave the pink filtrate undisturbed. Pink crystals will be formed after 3 days and dry the crystals in air. IR (KBr, cm^{-1}): 3063 (s), 1614 (s), 1568 (s), 1537 (s), 1487 (m), 1403 (s), 1219 (m), 1070 (m). Magnetic moment: 6.54 BM at 25°C; UV-vis: λ_{max} (MeOH) 519 nm, $\varepsilon = 40.10$ M^{-1}cm^{-1}.

Synthesis of [Co$_2$(μ-H$_2$O)(μ-OBz)$_2$(OBz)$_2$(Py)$_4$] 1.5(C$_6$H$_6$)

Take finely ground sodium benzoate (0.29 g, 2 mmol) and cobalt(II) chloride hexahydrate (0.238 g, 1 mmol) and heat at 100 °C for 15 min, during which the pink colour becomes blue. Cool the mixture to room temperature and transfer it into a round bottom flask. To this mass, add benzene (20 ml) and stir at room temp for 5 min followed by the addition of pyridine (0.158 g, 2 mmol). The blue supernatant liquid turns pink. Filter the residue and keep the pink filtrate undisturbed. Pink crystals will be obtained after 3 days. IR (KBr, cm^{-1}): 3446 (b), 3066 (bs), 2361 (b), 1627 (s), 1601 (s), 1573 (s), 1537 (m), 1485 (m), 1393 (s), 1068 (m), 1038 (m). Magnetic moment: 6.52 BM at 25°C; UV-vis: λ_{max}(MeOH) 519 nm, $\varepsilon = 40.32$ M^{-1}cm^{-1}.

Synthesis of aqua-*bis*pyridine-di-benzoato zinc(II)

Add cobalt(II) acetate tetrahydrate (0.25 g, 1 mmol) to a well stirred solution of benzoic acid (0.24 g, 2 mmol) in methanol (20 ml) and stir it for 12 hrs at room temperature. After that add pyridine (0.16 g, 2 mmol) at room temperature and stir it for half an hour. On standing pink crystals will be formed after 6 days. Dry the crystals in air. IR (KBr, cm^{-1}): 3265 (wb), 3058 (w), 1593 (s), 1540 (s), 1572 (m), 1485 (w), 1445 (m), 1408 (s), 1374 (s), 1065 (m), 1039 (m). Magnetic moment (27^0 C) 4.04 B.M. UV-vis: λ_{max} (MeOH) 517 nm, $\varepsilon= 25.07$ m^{-1}cm^{-1}.

Five coordinated zinc complex

Zinc(II) ions generally form six coordinated complexes. However, five coordinated complexes of zinc (II) can be prepared by controlling steric factors or by putting other constrains such as hydrogen bonding. One such five coordinate zinc(II) complex can be obtained from the reaction of zinc(II) acetate with benzoic acid in the presence of 8-aminoquinoline (equation 3.32).

$$\ldots\ldots(3.32)$$

Procedure for preparation of five coordinated zinc complex

Stir a solution of zinc(II) acetate dihydrate (0.22 g, 1 mmol) with benzoic acid (0.24 g, 2 mmol) and 8-aminoquinoline (0.29 g, 2 mmol) in methanol (30 ml) at room temperature for four hours. On standing the solution will give colorless crystals after two days. IR (KBr, cm^{-1}) 3467 (w), 3263 (m), 3078 (w), 2786 (w), 2320 (w), 1921 (w), 1870 (w), 1792 (w), 1629(s), 1511(m), 1470 (m),1368 (s), 1265 (w), 1148 (m), 1071(s), 1025 (m). ^{1}HNMR (CD$_3$OD, 400MHz, ppm) 8.92 (s, 4H), 8.32 (d, J = 9Hz, 2H), 7.94 (dm, J = 9Hz, 4H), 7.55-7.46 (m, 8H), 7.40-7.29 (m, 6H).

Trinuclear manganese(II) complex

Manganese(II) acetate tetrahydrate, 2-nitrobenzoic acid and 8-aminoquinoline (AQ) forms a trinuclear manganese(II) complex. The structure of trinuclear complex obtained from this reaction is shown in figure 3.28.

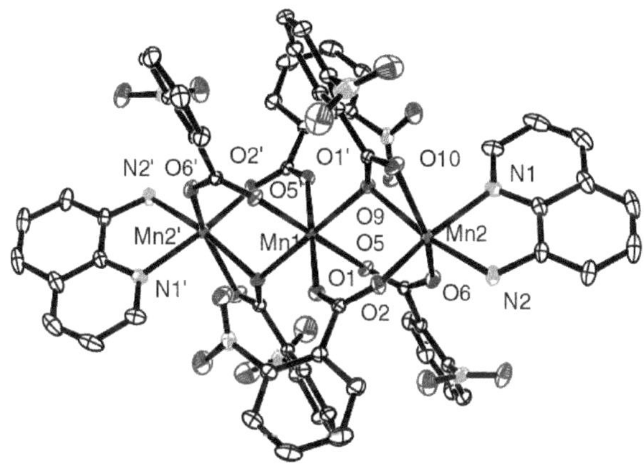

Fig. 3.28 A trinuclear manganese complex [Mn$_3$(2-NO$_2$C$_6$H$_4$COO)$_6$(AQ)$_2$] (hydrogen atoms are omitted for clarity of picture)

Synthesis of trinuclear complex [Mn$_3$(2-NO$_2$C$_6$H$_4$COO)$_6$(AQ)$_2$]

Mix manganese(II) acetate tetrahydrate (0.25 g, 1 mmol) and *o*-nitrobenzoic acid (0..33 g, 2 mmol) in a mortar and heat the mixture at 80 °C in an oven for 20 minutes. Cool the mixture and add slowly a methanolic solution of 8-aminoquinoline (0.29 g, 2 mmol in 10 ml) and stir for five minutes. To this solution add toluene (5 ml) and make the solution completely homogeneous. Leave the green colored solution for crystallization. Dark brown colored crystals appears on standing. IR (KBr, cm^{-1}): 3293 (bw), 1624 (s), 1527 (s), 1378 (s), 1143 (w), 1076 (w). Molar Conductance: 118.0 mol^{-1}m^{2}S in methanol; 5.4 mol^{-1}m^{2}S in acetonitrile; 33.2 mol^{-1}m^{2}S in N,N′- dimethylformamide.

Ring opening reaction of cyclic anhydride

Ring opening reactions of an anhydride can be used to prepare metal carboxylate complexes. For example, chelated nickel complex can be prepared by ring opening reaction of 2,3-pyridine dicarboxylic anhydride (equation 3.33).

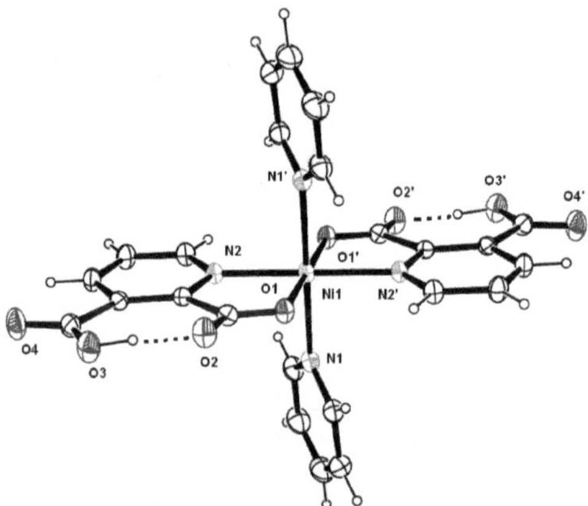

$$\ldots\ldots(3.33)$$

From this reaction a six coordinated nickel complex is prepared by adding pyridine (Fig. 3.29). Pyridine occupies the axial positions, where the two pyridine carboxylic acid ligands occupy the square planar positions. In the nickel complex there is a free carboxylic acid that is involved in intra-molecular hydrogen bonding.

Fig. 3.29 Structure of the complex $[Ni(LH)_2(H_2O)_2(py)_2]$ (LH_2 = 2,3-pyridine dicarboxylic acid py = pyridine) Some selected bond distances (Å) and angles (°) of the complex are Ni1−N2, 2.05; Ni1−N1, 2.13; Ni1−O1, 2.03; <N1−Ni1−O1, 89.21; <N2−Ni1−O1, 100.89; <N1−Ni1−N2, 90.80.

Synthesis of $[Ni(LH)_2(H_2O)_2(py)_2]$

To a solution of 2,3-pyridine dicarboxylic anhydride (0.3 g, 2 mmol) dissolved in methanol (10 ml, at 60 °C) add a solution of nickel(II) acetate tetrahydrate (0.25 g, 1 mmol) in methanol (10 ml). Stir the reaction mixture at room temperature for half an hour and to this mixture add pyridine (py) (0.16 g, 2 mmol). The resulting solution on standing will give pink crystal. IR(KBr, cm^{-1}) 3414 (bs), 1712 (m), 1599 (m), 1559 (s), 1480(s), 1441(s), 1418 (m), 1369 (s), 1275 (s), 1210 (s), 1152 (m), 1135 (m), 1101(s), 1062 (m).

Chelating carboxylate complex

An example of a chelating carboxylate is (2,2′-bipyridine)dibenzoato nickel(II);   where 2,2′-bipyridine acts as a chelating ligand. In this complex the two benzoate groups are chelating and the complex has a distorted octahedral geometry Fig. 3.30.

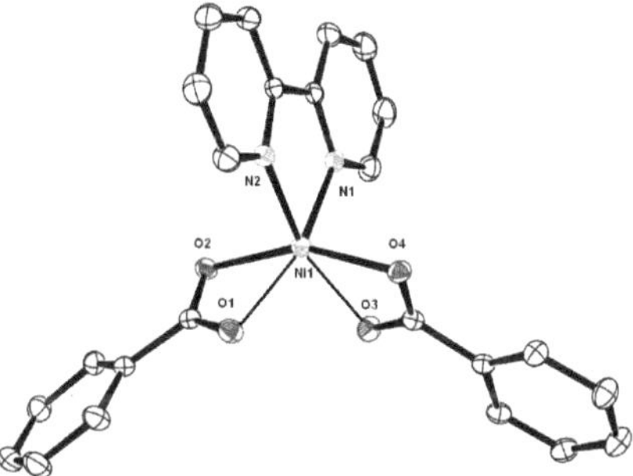

Fig. 3.30 Structure of (2,2′-bipyridine)dibenzoato nickel(II) (hydrogen atoms are omitted for clarity). Some bond distances (Å) and angles (°) are Ni1−N2, 2.03; Ni1−N1, 2.05; Ni1−O4, 2.05; Ni1−O2, 2.07; Ni1−O1, 2.15; Ni1−O3, 2.15; <N2−Ni1−N1, 79.48; <N2−Ni1−O4, 100.69; <N1−Ni−O4, 99.99; <N2−Ni1−O2, 96.40; <N1−Ni1−O2, 98.95; <N2−Ni1−O1, 96.26; <O4−Ni1−O1, 99.35; <N1−Ni1−O3, 97.57; <O1−Ni1−O3, 91.91.

Acetylacetonato complexes

Acetylacetone is an active methylene compound; it exists in keto-enol form (scheme 3.2). The enol form can form delocalized structure and is a good chelating ligand. Due to the ability of acetylacetone for formation of

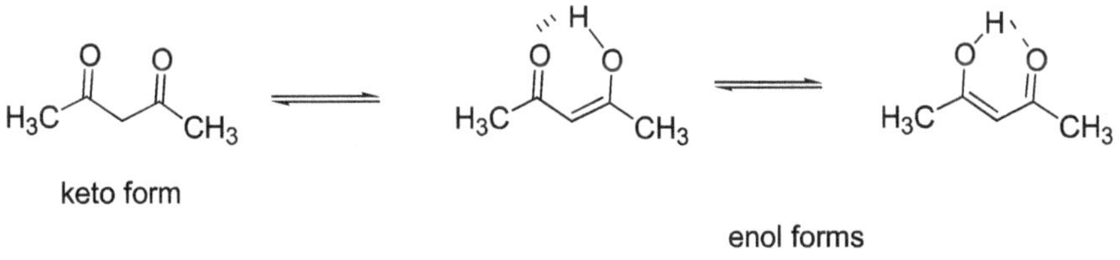

Scheme 3.2 Keto-enol form of acetylacetone

chelate complexes, acetylacetone forms varieties of stable metal complexes. The acetylacetonato complexes of metals are interesting as they serve precursor for metallo-organic chemical vapor deposition, for preparation of films or deposition at specific sites of metal oxides and metal itself. The metal complexes of acetylacetone are also soluble in varieties of organic solvents and thus this class of inorganic compounds is good for studies of their reactivities in different solvents. The acetylacetone complexes behave like aromatic systems and they undergo electrophilic substitution reactions at the active methylene carbon. In this regard a good amount of substitution reactions with kinetically inert acetylacetonato cobalt(III) complexes are carried out. Some of the reactions are shown in scheme 3.3.

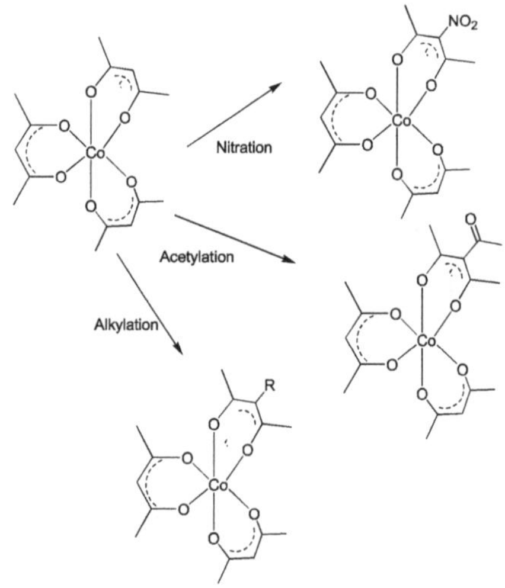

Scheme 3.3 Substitution reactions on delocalised ring of metal acetylacetonato complexes

The complex $[Eu\{(OCC(CH_3)_3CHCOCF_2CF_2CF_3)\}_3]$, known as $Eu(fod)_3$ is used in NMR spectroscopy as a NMR shift reagent. Originally $Eu(fod)_3$ was used for analyzing diastereomeric compounds by NMR spectroscopy (Fig. 3.31). The paramagnetic Eu^{3+} induces chemical shifts of the protons near a Lewis basic site(s) of a molecule. This principle is used to resolve closely-spaced signals of hydrocarbons and related compounds. A small amount of shift reagents are used for this purpose as the excessive use of the complex causes NMR line broadening.

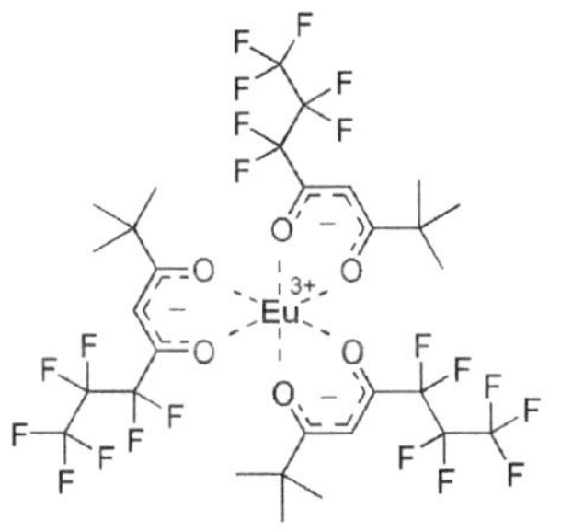

Fig. 3.31 Structure of Eu(fod)$_3$

Eu(fod)$_3$ is also a catalyst for stereoselective catalysts in reactions, such as, Diels-Alder (4+2 cycloaddition) reactions and in Aldol condensation reaction between carbonyl compounds. Acetylacetonato complexes of lanthanides are used as ^{1}HNMR shift reagent. The lanthanide complexes of acetylacetone derivatives with chiral centers are used in NMR spectroscopy for determination of enantiomeric excess. Some of the vanadium acetylacetonato complexes show interesting *cis-trans* equilibrium as illustrated in scheme 3.4

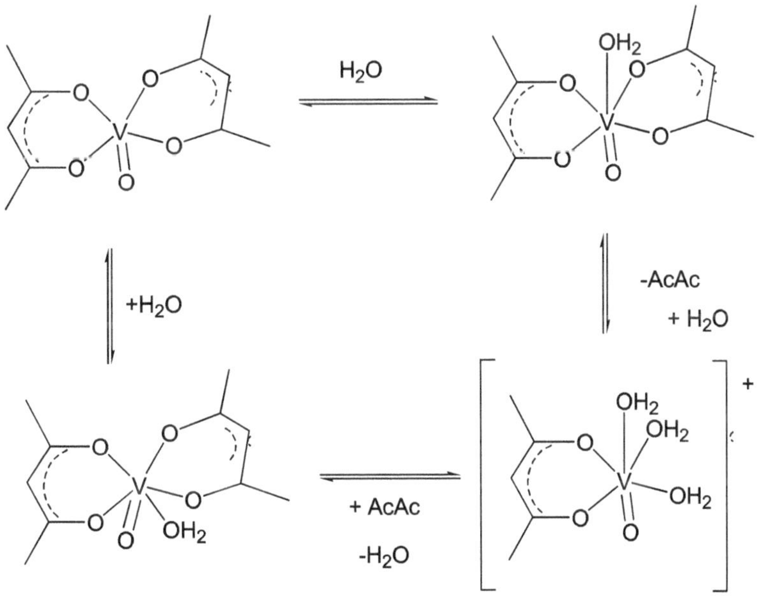

Cis –trans equilibriation by water in vanadium acetylacetonato complexes

Scheme 3.4

Such equilibrium are easily monitored in solution by recording electron spin resonance spectra of the complexes with change in concentration of water. The major product is the complex with acetylacetonato group *trans* to the oxo group, whereas in the minor complex has the solvent *cis* to the oxo group. This causes a minor distortion of the vanadium out of the plane toward the oxo group. This change in geometry is reflected in the esr spectra. The vanadium at +4 oxidation state is a d^1 system and the vanadium has $I = 7/2$; which results in $2nI + 1 = 8$ numbers of signals due to hyperfine splitting.

Tris-acetylacetonato iron(III) is a neutral complex and it is soluble in various organic solvents like chloroform, acetonitrile etc. It has three acetylacetonato ligands occupying the six co-ordination positions of a distorted octahedral geometry shown in Fig. 3.32.

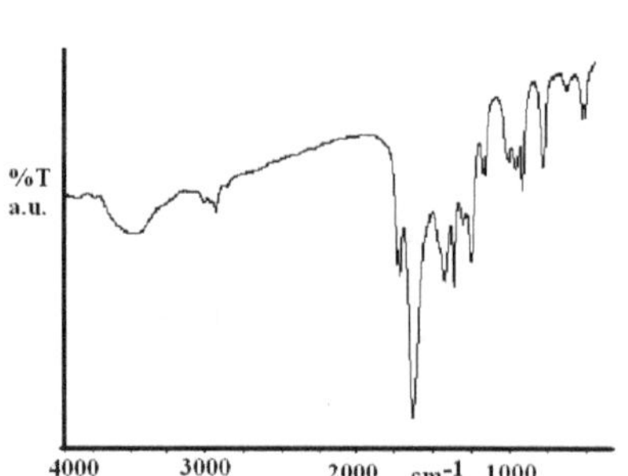

Fig. 3.32 Structure of *tris*-acetylacetonato iron(III)

The complex *tris*-acetylacetonato iron(III) has IR absorbances due to ν_{OH} at 3425 cm^{-1}, $\nu_{C=C}$ and $\nu_{C=O}$ at 1572 cm^{-1} and 1525 cm^{-1}; it has also absorptions due to $\delta_{s(CH3)}$; at 1366 cm^{-1}, and for ν_{C-C} at 1270 cm^{-1}. For comparison purpose the IR spectra of the parent compound namely acetylacetone is given in Fig. 3.33.

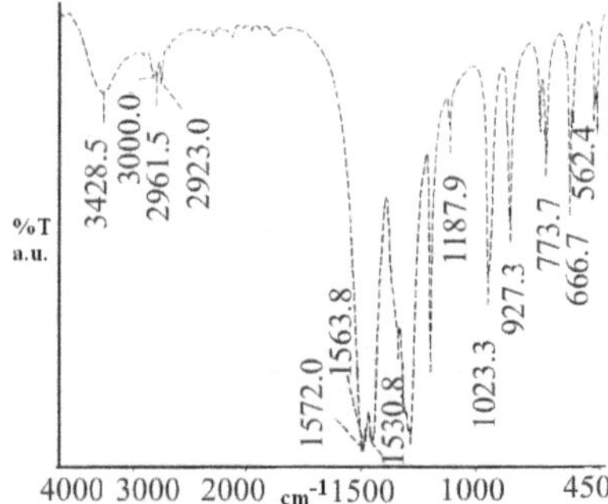

Fig. 3.33 FT-IR spectra of (left) acetylacetone (neat);on right *tris*-acetylacetonato iron(III) (KBr, cm^{-1})

One of the way to prepare the *tris*-acetylacetonato iron(III) complex is is by the following scheme:

$$FeCl_3 + 3KOH \rightarrow Fe(OH)_3 + 3KCl \qquad .....(3.34)$$

$$Fe(OH)_3 + 3\,AcAcH \rightarrow Fe(AcAc)_3 + 3H_2O$$

where   AcAcH = acetylacetone   .....(3.35)

Preparation of *tris*-acetylacetonato iron(III)

Dissolve anhydrous iron(III) chloride (1.5 g, 9.25 mmol) in water (20 ml) in a 100 ml beaker. Add an aqueous solution of potassium hydroxide (20% w/v) in parts with constant stirring to precipitate the iron (III) as ferric hydroxide. Continue the addition of alkali till the pH of the solution rises to 8. Allow the suspended precipitate to settle down at the bottom of the beaker; by doing this the supernatant liquid becomes colorless. Wash the flocculants several times with water by decantation, finally filter the residue through filter paper. Wash the residue twice again with cold water. Transfer the precipitate into a 100 ml beaker containing water (10 ml). Add distilled acetylacetone (3 ml, 29.5 mmol) to the slurry and mix them thoroughly. An exothermic reaction leads to the formation of deep red shiny crystals of *tris*-acetylacetonato iron(III). Allow the mixture to attain room temperature. Place the reaction container in an ice-water bath for 15 min. Filter the compound through filter paper and dry in vacuum.

Schiff base complexes

Schiff base finds an important place in coordination chemistry. They are routinely used to build new structures and also to make molecules that mimic biological activity. Simple and commonly encountered reactions in coordination chemistry to design ligand, involve condensation reactions of nitrogen containing compounds with carbonyl compounds. The hydroxylamine and hydrazine salts condense easily with carbonyl compounds and forms imines and related compounds which serves as lignads in inorganic chemistry. The oximes, formed from hydroxylamine with aldehydes are also used in functional group transformations. Functional groups such as cyanide and amide are can be easily obtained from oximes (scheme 3.5). The hydrazine derivatives are used as characteristic test for carbonyl functional groups and also as reference material for their characterization. Derivatives of primary amines on condensation with carbonyl compounds gives C==N bonded compounds known as imines. When these imines are used in coordination chemistry they are known as Schiff bases and the complexes derived from them are known as Schiff base complexes. The use of Schiff bases in inorganic chemistry is very large and they have significantly contributed to understanding of basic principles of coordination chemistry.

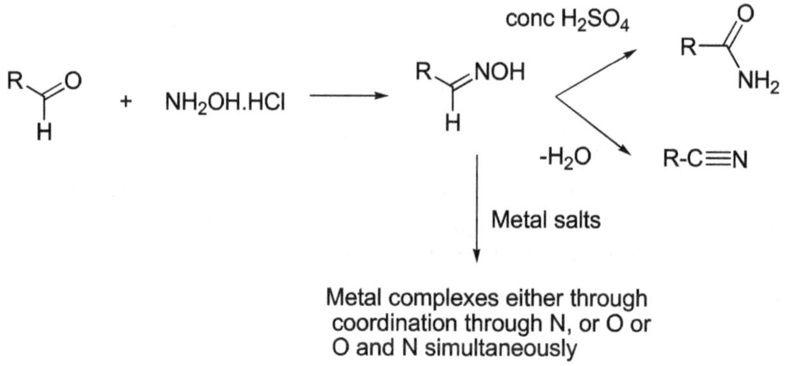

Scheme 3.5 Various reactions of oximes

The imines with additional binding sites are very useful ligands. The condensation reactions of salicyldehyde with amines lead to imine derivatives which are very good chelating ligands as illustrated in scheme 3.6.

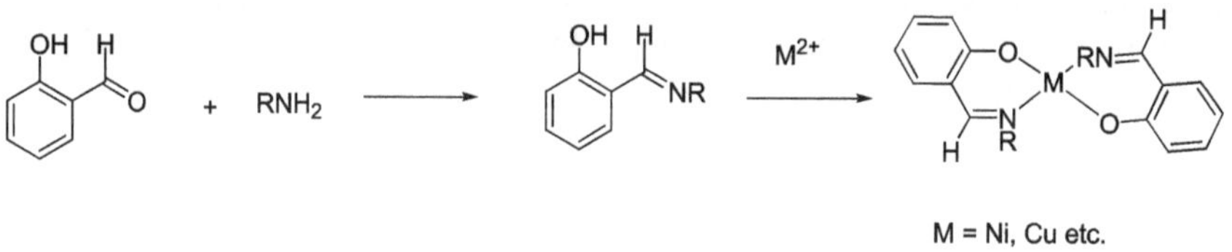

Scheme 3.6 Metal Schiff base complexes

The complexity in ligation of Schiff base can be increased further by adding more number of binding sites to such ligands. Thus a ligand can be designed to make polynuclear metal complexes. For example condensation of ethylenediamine or similar di or tri amine with salicyldehyde can provide multiple chelation centers. For example, condensation of salicyldehyde with ethylenediamine leads to bis-(N,N-disalicylidene ethylenediamine), which has two chelation sites within the same ligand.

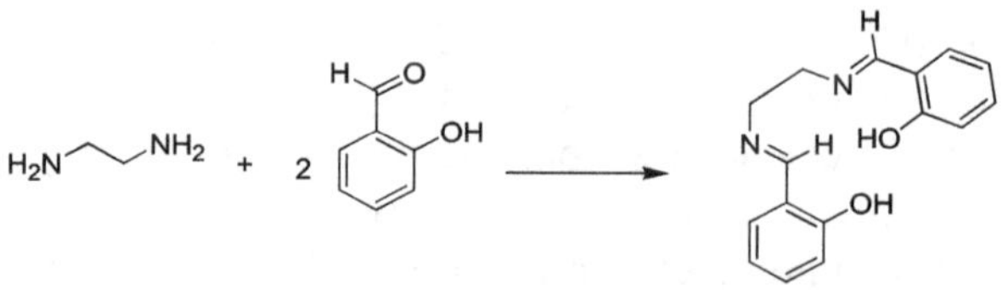

$$.....(3.36)$$

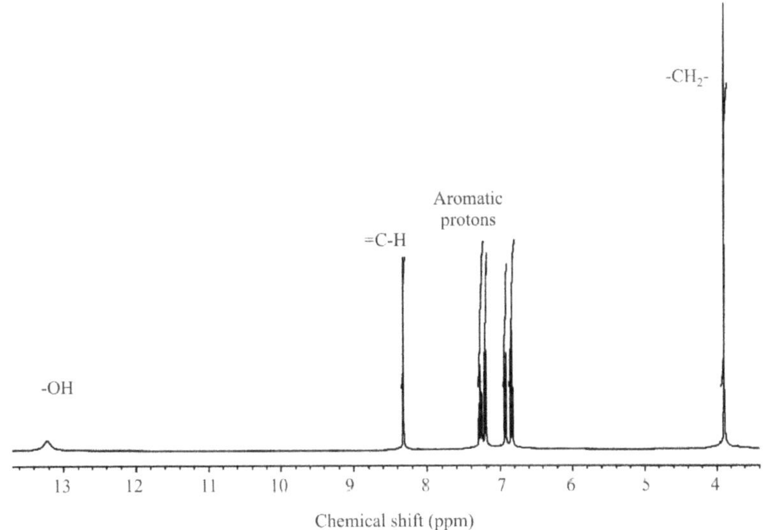

Fig. 3.34 ¹HNMR spectra (400 MHz, CDCl₃, ppm) of bis-(N,N-disalicylidene ethylenediamine)

The ^{1}HNMR spectra of *bis*-(N,N-disalicylidene ethylenediamine) has the signals for the CH₂ groups at 3.9 ppm. The aromatic peaks appear as A₂B₂ pattern at 6.9 and 7.2 ppm. The methine proton i.e. the proton flanked by the imine nitrogen and aromatic ring appears at 8.3 ppm whereas the OH appears at 13.2 ppm (Fig. 3.34).

The *bis*-(N,N-disalicylidene ethylenediamine) complex with two cobalt(II) centers to form complex *bis*-(N, N' disalicylalethylene diamine)-μ-aqua-di-cobalt(II) is based on the following equation:

.....(3.37)

This complex is an example of an aqua bridged cobalt(II)complex. There are other examples of cobalt complexes in which the same ligand binds to two metals when two chelating sites exist within the same ligand. The ligand has the $\nu_{C=N}$ and $\nu_{C=C}$ frequencies at 1634 cm^{-1} and 1600 cm^{-1}

respectively in its IR spectra. Whereas in the complex the peak at 1600 cm^{-1} due $\nu_{C=C}$ is intact but the $\nu_{C=N}$ merges with the $\nu_{C=C}$ stretching frequency. For comparison purpose the IR spectra of the ligand and the cobalt complex are given figure in Fig 3.34 and 3.35 respectively.

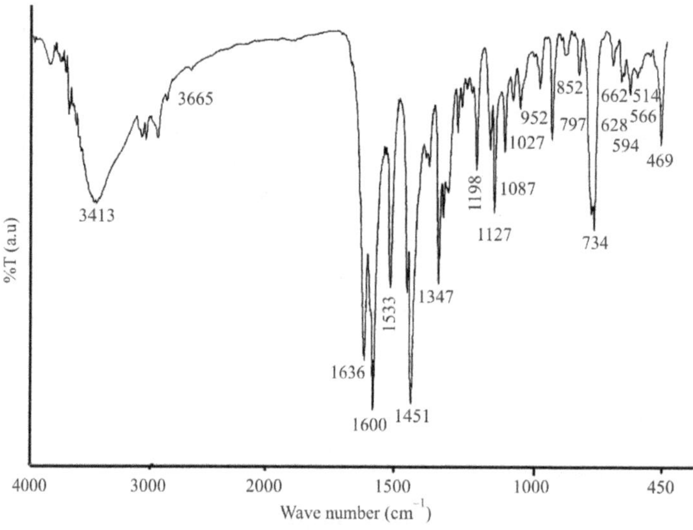

Bis-(N,N-disalicylidene ethylenediamine)

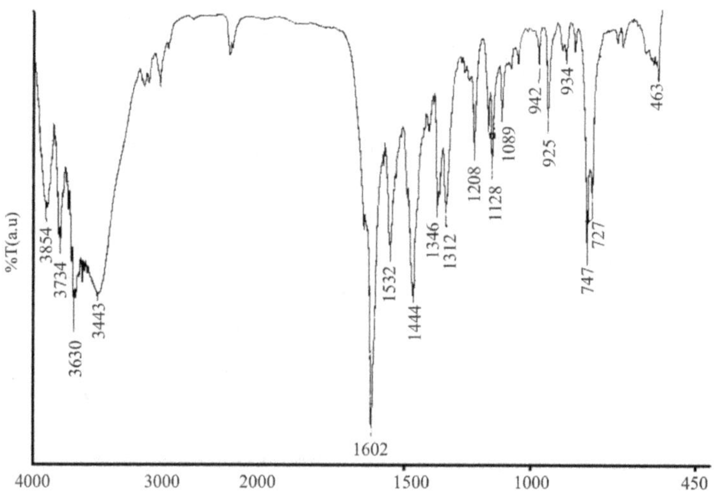

Bis-(N,N'-disalicylalidene ethylenediamine)-μ-aqua- di-cobalt(II)

Fig. 3.35 FT-IR spectra (KBr, cm^{-1})

Preparation of *bis*-(N,N′-disalicylidene ethylenediamine)

Dissolve salicylaldehyde (3.05 g, 0.03 mol) in ethanol (25 ml) in a round bottom flask. Heat the solution to ~70 °C and add a solution of ethylenediamine (0.75 g, 0.013 mol) dissolved in ethanol (3 ml). Heat the resulting yellow solution under reflux for 30 mins. Reduce the volume of the solution to half by evaporating the solvent and after cooling, filter the yellow crystalline precipitate under suction and wash with ice-cold ethanol. Recrystallize the product from methanol and dry at room temperature under vacuum.

Preparation of *bis*(N, N′-disalicylidene ethylenediamine)-μ- aqua-di-cobalt(II)

Dissolve finely ground N,N′-disalicylidene ethylenediamine (1.34 g) in water (150 ml) and to this solution add sodium hydroxide (3.9 g) and sodium acetate trihydrate (5 g). Continue stirring for 10-15 minutes. While stirring is continued add a solution of cobalt(II) chloride hexahydrate (1.23 g) dissolved in hot water (25 ml). Stir the reaction mixture until it turns to a reddish-brown paste. Allow the reaction mixture to stand for at least 15 minutes. Centrifuge the reaction mixture until most of the mother liquor has been removed and a hard cake remains. Wash the precipitate three times with water (each time with about 10 ml of water). Remove the cake from the centrifuge tube and mix thoroughly with water (75 ml) so that no large particles remain and uniform slurry is obtained. The cake is further centrifuged as dry as possible. It is then broken into small pieces and dried at 100 °C under reduced pressure.

Estimation of cobalt

Take an accurately weighed amount (range 0.1 to 0.2 g) of the cobalt(II) complex in minimum amount of hot dilute hydrochloric acid (1:1). Treat the solution with sodium hydroxide (5% w/v) in order to precipitate cobalt(II) hydroxide. Use of excess alkali should be avoided. Centrifuge the precipitate and wash carefully with distilled water. Filter the precipitate and oxidize it to cobalt(III) by boiling with a small amount of hydrogen peroxide (3 ml, 30% v/v in 20ml water) solution. Remove the excess peroxide by boiling for nearly 30 min. Dissolve the precipitate in warm acetic acid and dilute it to 200 ml with water. Add 1-nitroso-2-

naphthol reagent dropwise to the warm solution with stirring. A red-brown precipitate of $[Co(C_{10}H_6O_2N)_3]$ will appear. Filter the precipitate through a pre-weighed sintered glass crucible and dry in an oven at ~110 °C and find out the constant weight i.e. repeat the weighing, until the value is constant. Find out the percentage of cobalt in the complex.

Polyoxo metallates

Poly-oxo compounds finds important place in inorganic chemistry. They provide structural information on many naturally occurring samples such as rocks, clay and mineral. They form various structures with M-O bonds. These compounds are useful as porous materials and used in molecular recognition and as molecular sieves. The formation of different types of polyoxo species depends on the stoichiometry, reaction conditions etc. In some cases like vanadium, different oxo-species has different colors and these oxo species can be identified by different color changes at different reaction conditions.

Different vanadium complexes can be prepared from same ligand at different pH; for example, some vanadium complexes derived from 8-hydroxyquinoline ligand are prepared just by varying pH. Some such species are shown in Fig. 3.36.

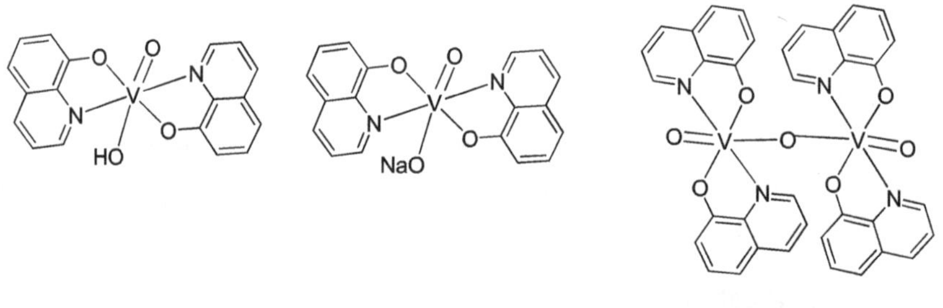

Fig. 3.36 Various vanadium oxo-complexes with anion of 8-hydroxyquinoline

Among the various metals molybdenum, tungsten, zirconium polyoxo metallates have diverse structures and such structural chemistry are well established. Molybdenum oxo-species have different structures depending on pH of medium from which they are prepared. Depending on pH following equilibrium exists (scheme 3.7):

$$7\,[MoO_4]^{2-} + 8\,H^+ \rightleftharpoons [Mo_7O_{24}]^{6-} + 4H_2O$$

$$8\,[MoO_4]^{2-} + 12\,H^+ \rightleftharpoons [Mo_8O_{26}]^{4-} + 6H_2O$$

$$36\,[MoO_4]^{2-} + 64\,H^+ \rightleftharpoons [Mo_{36}O_{112}]^{8-} + 32H_2O$$

Scheme 3.7

In addition to these, by carefully adjusting pH other species such as $[Mo_6O_{19}]^{20-}$, $[Mo_{10}O_{34}]^{8-}$ etc. are prepared. Two examples of molybdenum poly oxometallate namely a hexamolybdate and an octamolybdate are synthesised by simple control of stoichiometry and acidity of the medium are given below (Scheme 3.8):

$$6\,Na_2MoO_4 + 10\,HCl + 2(n–C_4H_9)_4NBr$$
$$\longrightarrow [(n–C_4H_9)_4N]_2[Mo_6O_{19}] + 10\,NaCl + 2\,NaBr + 5\,H_2O$$
$$8Na_2MoO_4 + 12HCl + 4(n–C_4H_9)_4NBr$$
$$\longrightarrow [\beta–(n–C_4H_9)_4N]_4[Mo_8O_{26}] + 12NaCl + 4NaBr + 6H_2O$$

Scheme 3.8

Distinction of these complexes can be made by analyzing the IR spectra of each of these complexes. In the Fig. 3.37 the distinction among the tetra-butylammonium hexamolybdate (VI) and tetra-butylammonium octamolybdate (VI) can be seen. In the two IR spectra similar features above 800 cm^{-1} are seen due to the presence of same cation namely tetrabutylammoium cation in both the complexes. However, in the far-IR features (below 800 cm^{-1}) the absorptions are different; which arises due to Mo-O bond stretching and shows the feature of two different cores. It is practically impossible to correctly predict the unknown structures of these oxo-clusters by simple IR spectroscopy. The structures are resolved in conjuncture with multiple numbers of data from different spectroscopic techniques along with the determination of metal content in such complexes and also by determination of amount of other related elements present in the complex. Sometimes chemical reactivity of such clusters also helps in ascertaining structures. Finally, the X-ray tools like power X-ray diffraction patterns and single crystal structure determination are used for ascertaining such structures.

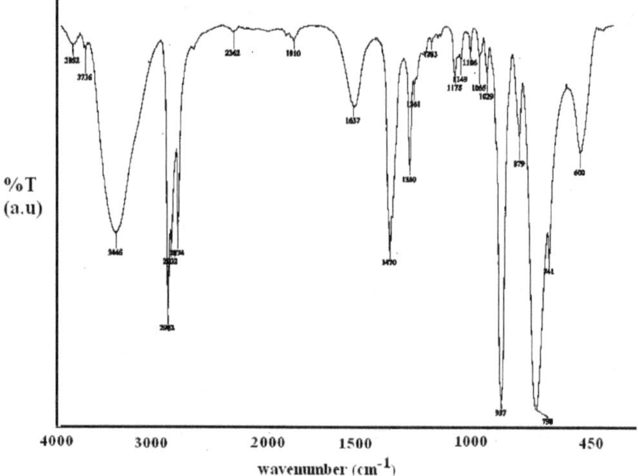

Tetra-butylammonium hexamolybdate (VI)

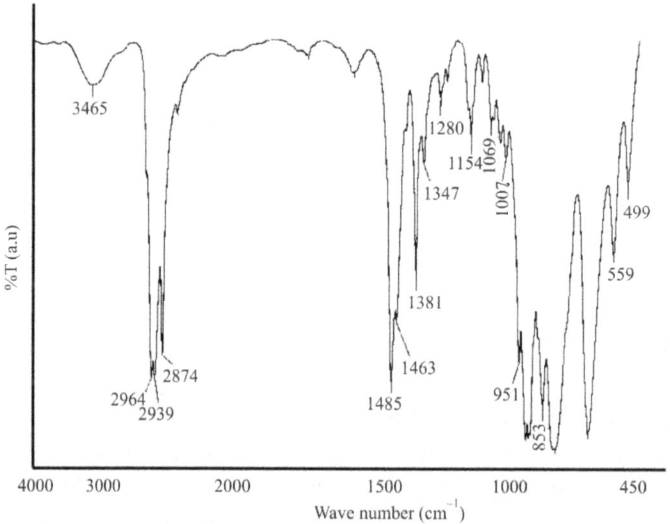

Tetra-butylammonium octamolybdate (VI)

Fig. 3.37 FT-IR spectra (KBr, cm^{-1})

Preparation of tetrabutylammonium hexamolybdate (VI)

Take a solution of sodium molybdate dihydrate (2.5 g, 10.3 mmol) in water (10 ml). Acidify it with hydrochloric acid (2.9 ml, 17.4 mmol) in a conical flask (50 ml) with vigorous stirring over a period of one minute at room temperature. Add a solution of tetrabutylammonium bromide (1.2

g, 3.75 mmol) in water (2 ml) with vigorous stirring to form a white precipitate. Heat the resulting slurry to 80 °C with stirring for 45 mins. During this period the white solid changes to yellow. Collect the crude product with suction filtration and wash it with water (20 ml). Crystallize the product from acetone. After 24 hrs collect the yellow crystalline hexamolybdate by suction filtration. Wash the product with diethylether and dry in vacuum.

Preparation of tetrabutylammonium octamolybdate (VI)

Acidify a solution of sodium molybdate dihydrate (2.50 g, 10.4 mmol) in water (6 ml) with hydrochloric acid (6N, 2.6 ml, 15 mmol) in an Erlenmeyer flask. Add a solution of tetrabutylammonium bromide (1.7 g, 5.2 mmol) in water (10 ml) with vigorous stirring to cause immediate formation of a white precipitate. After stirring for 10 minutes, collect the precipitate by filtration with suction, and wash successively with water (10 ml), ethanol (10 ml), acetone (10 ml) and diethyl ether (10 ml). Dissolve this crude product in a minimum amount of acetonitrile and keep at 0 °C overnight to get colorless block-shaped crystals. Collect the crystals by suction filtration and dry for 12h in vacuum. The transparency of the crystals is lost upon drying.

Estimation of molybdenum

Neutralize an accurately weighed sample (0.15 g), in about 30 ml of water, and then acidify with a few drops of sulphuric acid (1M) (check with methyl rcd). Add ammonium acetate (2M, 5 ml), dilute the solution to 50-100 ml, and heat it to boiling. Precipitate the molybdenum by the addition of 3% solution of oxine in dilute acetic acid, until the supernatant liquid becomes yellow. Boil the solution gently and stir for about 5 minutes. Filter the solution through a sintered glass crucible of known weight, wash with hot water until free from the oxine, and dry to constant weight at 130-140 °C. Weigh the precipitate it as $MoO_2(C_9H_6ON)_2$, calculate the amount of molybdenum in the complex.

Peroxide complexes

Hydrogen peroxide is an important oxidizing agent. Its decomposition is catalysed by various metal ions. The hydrogen peroxide decomposes to oxygen and in this decomposition reaction water is the only side product which is environmentally benign. Thus, it is a very useful reagent in green chemistry. It can have different binding modes with metal ions. Some of the binding modes that that are observed in transition metal complexes are listed Fig. 3.38.

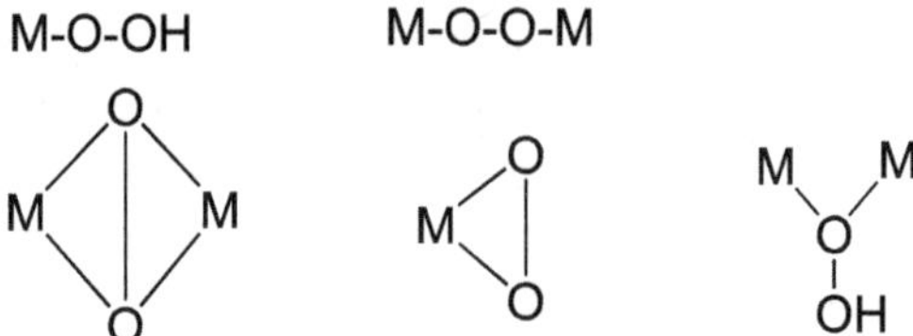

Fig. 3.38 Different types of binding modes of peroxide with metal ions

Borax reacts with hydrogen peroxide and forms peroxo compound which releases oxygen in a controlled manner. This process is useful for cleaning purpose and is used in washing powder for better cleaning action.

$$Na_2B_4O_7 + 4\ H_2O_2 + 2\ NaOH \rightarrow 2\ Na_2B_2O_4(OH)_4 + H_2O \ .....(3.38)$$

Similarly, urea forms inclusion complex with hydrogen peroxide. In such inclusion complex the hydrogen bonded structures are formed and the hydrogen bonded inclusion complexes are obtained as solids. These inclusion complexes also control the release of oxygen from hydrogen peroxide and provides method for generation of pure hydrogen peroxide.

Bleaching action of peroxoborate

Peroxoborates are good bleaching agents than hydrogen peroxide. The electrophilic nature of the peroxy group increases on coordination to the boron centre. Thus, sodium perborate is a superior bleaching agent to hydrogen peroxide, but efficiency of former is higher at elevated temperature. The process of release of hydrogen peroxide in cleaning action can be represented by the following scheme 3.9.

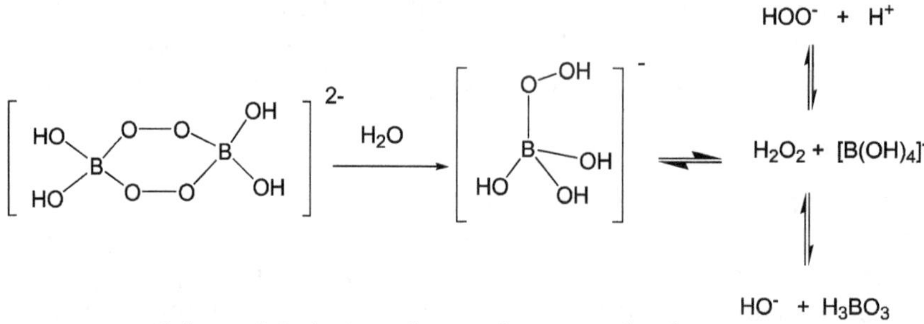

Scheme 3.9 Action of peroxoborate as cleaning agent

Auto-oxidation of 2-ehthyl-9,10-dihydroxyanthracene gives 2-ehtylanthraquinone and when this reaction is carried out in the presence of oxygen, the oxygen gets converted to hydrogen peroxide. This method is industrially used for synthesis of hydrogen peroxide.

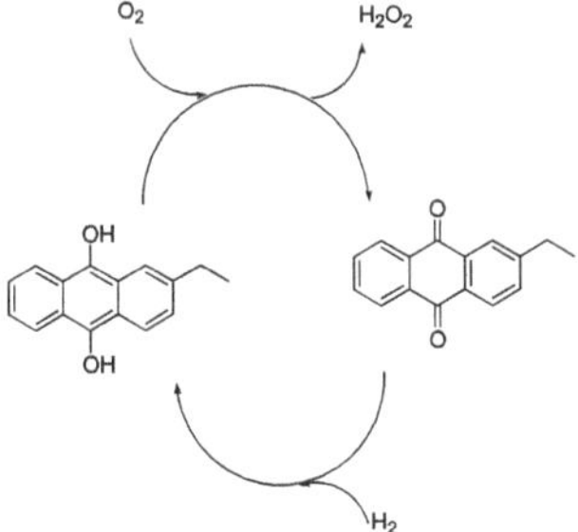

Scheme 10 Production of hydrogen peroxide

In this reaction, the hydroxy groups are deprotonated and oxidised to quinonic compound. The anthraquinone derivative is to be reduced back to the dihydroxy compound to make a catalytic cycle. For this purpose hydrogen gas is used in the presence of a metal catalyst. The overall reaction is:

$$H_2 + O_2 \rightarrow H_2O_2 \qquad\qquad .....(3.39)$$

Decomposition of hydrogen peroxide

Decomposition of hydrogen peroxide is catalysed by various catalysts. Determination of role of the metal complexes in such decomposition of H_2O_2 as well as determination of rate and understanding of mechanism of decomposition of hydrogen peroxide is important from biological point of view. For example, hydrogen peroxide can homolytically or heterolytically cleaved to generate hydroxyl or hydroperoxy radical. It may result in the formation of hydroxyl anion or peroxy dianion etc. These types of species are activated by metal to perform oxidation, hydroxylation. Such reactions also can lead to new peroxo bonded compounds. Kinetics of such reaction can be studied easily as the unreacted hydrogen peroxide during these reactions can be estimated by iodometric titration. However, there are many reactions of hydrogen peroxide that are caused by light are very fast and life time of the reactive species in such reactions are of the order of nano-second. In such cases, techniques like stop-flow are used for kinetic study.

Decomposition of hydrogen peroxide catalysed by ferric ions follows the kinetics of a first order reaction.

$$\overset{\text{FeCl}_3}{H_2O_2 \quad \rightarrow \quad H_2O + O_2} \qquad \dots(3.40)$$

The rate constant of the reaction is given by

$$\log\frac{a}{a-x} = \frac{k_1}{2.203}t \qquad \dots(3.41)$$

where, a is the initial concentration of hydrogen peroxide and x is the concentration of hydrogen peroxide consumed at time t. The progress of the decomposition can be followed by titrating the peroxide present by permanganate solution. Let V_0 and V_t be the volume of permanganate required to reduce hydrogen peroxide at the beginning and at any time t, then,

Initial amount of H_2O_2 a ∞ V_0

And the amount of H_2O_2 remaining at time t, $a - x \infty V_t$

$$\frac{a}{a-x} = \frac{V_0}{V_t} \qquad \dots(3.42)$$

and from equation 3.41

$$\log\frac{V_0}{V_t} = \frac{k_1}{2.303} \qquad \dots(3.43)$$

When $\log\dfrac{V_0}{V_t}$ is plotted against time, a straight line passing through the origin is obtained.

The slope of the straight line will give the value of $k_1/2.303$, thus k_1 can be calculated.

The reaction used for the estimation of hydrogen peroxide is

$$KMnO_4 + H_2SO_4 + H_2O_2 \quad \rightarrow \quad MnSO_4 + K_2SO_4 + O_2 \qquad \dots(3.44)$$

Potassium permanganate is colored and the manganese(II) sulphate is colorless, thus the end point of the reaction can be easily detected by change of color from violet to colorless.

Kinetics of ferric chloride catalyzed decomposition of hydrogen peroxide

Take a solution of hydrogen peroxide (100 ml, 2 volume) in a conical flask (250 ml) and place it on a thermostat. Make a ferric chloride solution (5% w/v, 10 ml) in water and place it on the thermostat. When they attain temperature 30°C, add ferric chloride solution (5 ml) to the hydrogen peroxide solution, and mix them well. Immediately withdraw 5 ml of the reaction mixture by a pipette and add it to ice cold solution of sulphuric acid (0.5N, 100 ml) taken in a conical flask (500 ml). Titrate the solution against potassium permanganate solution (N/20). This titre value corresponds to V_0. Repeat the process by withdrawing the reaction mixture and titrating against potassium permanganate solution at 5 minutes time intervals until the titre value becomes 40% of the initial value. These values correspond to V_t. Plot the values of log V_0/V_t against t. A straight line will be obtained. From the slope of the straight line calculate the value of k_1.

Potassium *bis*-(peroxo)-oxo-(1,10-phenanthroline)vanadium(V) trihydrate

The potassium *bis*-(peroxo)-oxo-(1,10-phenanthroline)vanadium(V) trihydrate is an example of a good oxidizing agent having bidentate peroxy ligand.

$$V_2O_5 \; + KOH \; + H_2O_2 \; + \quad \longrightarrow$$

$$.....(3.45)$$

This complex is prepared by reacting vanadium pentoxide with 1,10-phenanthroline and hydrogen peroxide in basic medium.

The IR spectra is shown in Fig. 3.39. The complex has IR absorbance at 3404 cm^{-1} due to ν_{OH} ; 1629 cm^{-1} due to δ_{H-OH}; 953 cm^{-1} due to $\nu_{V=O}$; at 830 cm^{-1} due to ν_{O-O} stretching frequencies. The thermogravimetry of the complex is shown in figure 3.40. It shows that the complex losses weight continuously on heating from room temperature to 600 °C to finally form nonvolatile K_2O and V_2O_5.

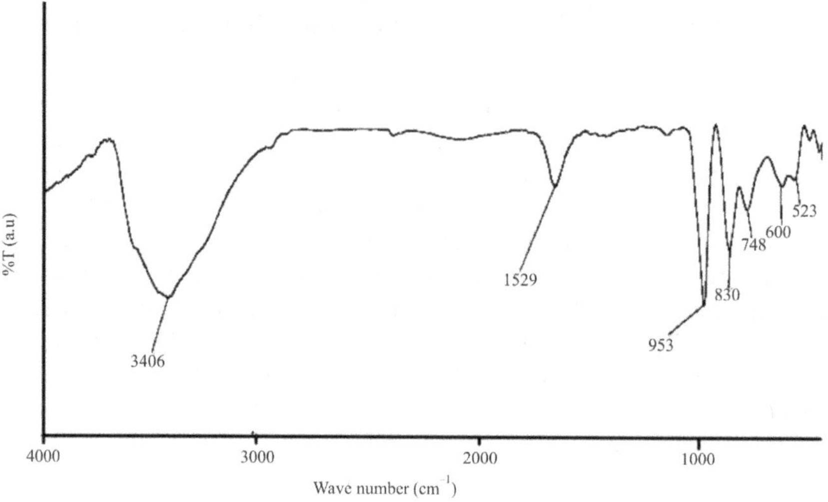

Fig. 3.39 FT-IR spectra (KBr, cm^{-1}) of *bis*-(peroxo)-oxo-(1,10-phenanthroline)vanadium(V) trihydrate

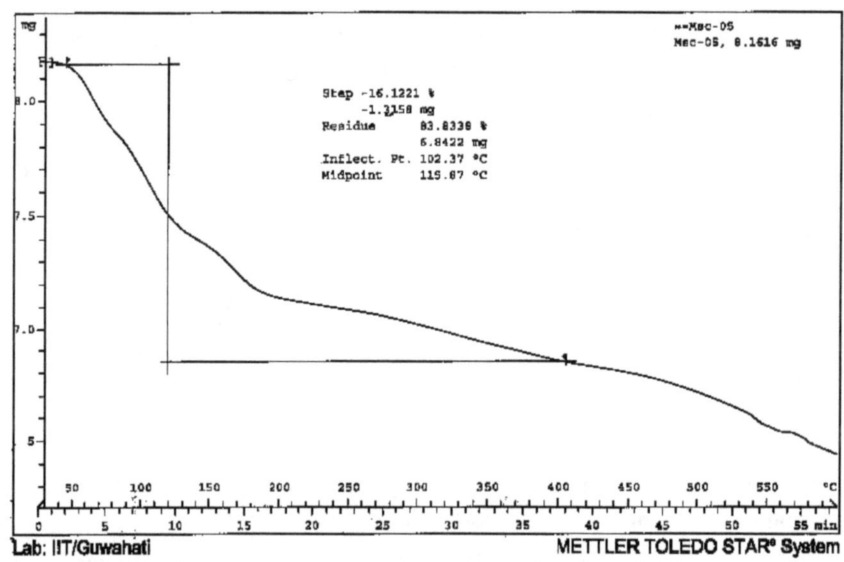

Fig. 3.40 Thermogram (heating rate 10°C/min) of potassium *bis*-(peroxo)-oxo-(1,10-phenanthroline)vanadium(V) trihydrate

Preparation of potassium *bis*-(peroxo)-oxo-(1, 10-phenanthroline) vanadium(V) trihydrate

Dissolve a mixture of vanadium pentoxide (0.91 g) and potassium hydroxide (0.65 g) in water (10 ml); add few drops of hydrogenperoxide solution. Cool the resultant clear solution over an ice-bath. Add hydrogenperoxide (30% v / v, 10 ml) drop wise to the ice-cold solution followed by of ethanolic solution of phenanthroline (1.6 g in 8 ml). Immediately crystals of the compound will appear. Wash the compound with ethanol. Dry it under vacuum.

Caution: Hydrogen peroxide in basic medium with organic substrates can cause explosion so handle under fume-cup board with shield.

Estimation of vanadium

Dissolve an accurately weighed sample (about 0.15 g) of the complex in distilled water (~100 ml) and boil the solution for 20 min. Cool the solution. Add ammonium persulphate (~1 g) and also add one or two drops of silver nitrate solution (1M). After addition of silver nitrate solution, boil the solution for about 30 mins. Cool the solution to room temperature. Estimate amount of vanadium by adding potassium iodide followed by titration of the liberated iodine with sodium thiosulphate.

Estimation of peroxide

Add a solution of boric acid (2 g, 32.3 mmol) in water (~100 ml) to a solution of accurately weighed sample (about 0.15 g) of the complex in sulphuric acid (25 ml, 5N). Titrate the solution with standard potassium permanganate (0.1N) solution.

Macrocylic ligands

Heteroatom such as oxygen, nitrogen or sulphur containing macrocycles are useful ligands for metal complexes. The complexes of macrocycles have extra stability. In this regard, the crown ethers are very important as they are capable of cation recognition. Analogous, cyclic nitrogen containing macrocyclic ligands are prepared from reaction of dicarbonyl compounds with diamines. These ligands are used to synthesize biological mimics. Poprphyrins are biologically important compounds. They are pyrole derivative, generally prepared by condensation reaction of pyrole with aldehydes.

The condensation reaction of ethylenediamine with acetone in presence of perchloric acid gives the perchlorate adduct of 5, 7, 7, 12, 14, 14-hexmethyl-1, 4, 8, 11-tetraazacyclo tetra deca-4, 1, 1-diene. This is an interesting macrocyclic compound and can form complex with different transition metal ions.

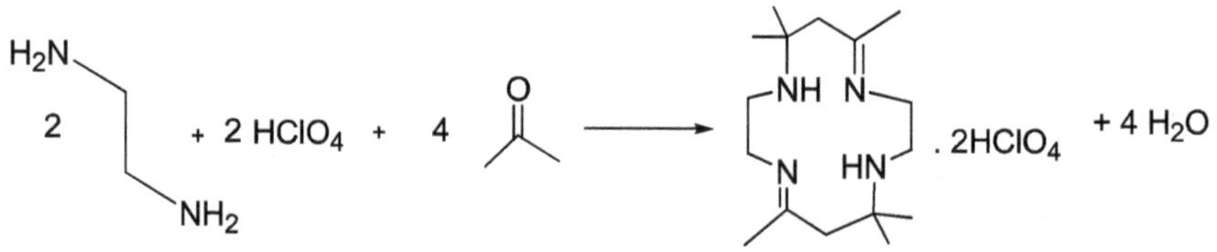

$$\text{.....(3.46)}$$

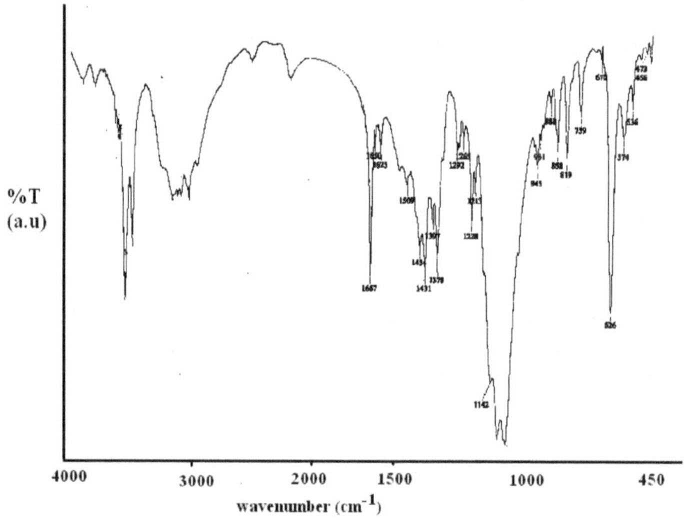

Fig. 3.41 FT-IR spectra (KBr, cm^{-1}) of perchlorate salt of 5, 7, 7, 12, 14, 14-hexmethyl-1, 4, 8, 11-tetraazacyclo tetra deca-4, 1, 1-diene

The presence of perchloric acid in the macrocycle is reflected in the observed in IR spectra of the compound. It has the characteristic strong absorption of perchlorate at 1125 cm^{-1}, the ν_{N-H} stretching absorption appears at 3465cm^{-1} and the $\nu_{C=N}$ is observed at 1667 cm^{-1}.

Synthesis and characterization of perchlorate salt of protonated 5, 7, 7, 12, 14, 14-hexmethyl-1, 4, 8, 11-tetraazacyclo tetra deca-4, 1, 1-diene

Take a solution of ethylenediamine (1 g, 0.33 mol) in dry acetone (25 ml) in a beaker (250 ml). Add a solution of perchloric acid (2.79 g, 60 %) dropwise to this solution with constant stirring. (Caution: The experiment should be carried out in hood. Use of organic substrate/ protic organic solvent with perchlorate may cause explosion). The solution becomes orange red. Once the addition of perchloric acid is completed, cool the reaction mixture to room temperature. A fine white crystalline product will be formed, filter the product. Wash the residue with acetone (5 ml) and dry in a dissicator over phosphorous pentoxide as desiccant.

The ligand forms copper(II) complexes on reaction with copper(II) perchlorate hexahydrate.

$$\dots\dots(3.47)$$

Synthesis of copper(II) complex of 5, 7, 7, 12, 14, 14-hexmethyl-1, 4, 8, 11-tetraazacyclo tetra deca-4, 1, 1-diene

Take the perchlorate salt of protonated 5,7,7,12,14,14-hexmethyl-1,4,8,11-tetraazacyclo tetra deca-4, 1,1-diene (10.1 mmol) in methanol (25 ml) in a beaker (100 ml). Neutralise the solution by adding sodium hydroxide (0.1M) and to this solution add a solution of copper(II) perchlorate hexahydrate (10 mmol) in methanol (10 ml) and stir at room temperature for 15 min. Concentrate the methanol solution by heating the solution over a water bath. On cooling violet crystalline complex is obtained. Filter the solid.

Reactions of trivalent phosphorus compounds

Phosphorus compounds undergo nucleophilic substitution reactions. Such substitution reactions are similar to carbon compounds in organic chemistry. However, phosphorus has multiple valencies +3 and +5; thus, the manipulation of phosphorus compounds requires extra care to avoid their inter-conversion during the course of a reaction. For example, the compounds of phosphorus at +3 oxidation state are easily oxidized to +5 oxidation state. Moreover, presence of the empty d-orbitals in phosphorus makes them hydrolytically susceptible. Thus, the reactions need to be performed under inert atmosphere. An example of substitution reactions on +3 phosphorus compound is the reaction of a secondary amine with phosphorous trichloride to form diamino substituted product.

$$PCl_3 + 2 \ (_iPr)_2NH \xrightarrow{\text{Toluene}} (_iPrN)_2PCl \qquad \dots\dots(3.48)$$

where $\qquad\qquad _iPr = $ isopropyl group

Preparation of *bis* (diisopropylamine)chloro phosphate

Take a solution phosphorus trichloride (1.92 g, 1.4 mmol) in dry toluene (20 ml) and in a two necked round bottom flask (100 ml) equipped with a reflux condensed and pressure equalizing dropping funnel (100 ml). The reflux condenser is to be attached to an out-let for continuous flow of nitrogen. Place a magnetic bar in the solution and flush a steam of nitrogen gas via side arm equipped with an inlet. Place diisopropylamine (6.09 g, 60 mmol) in toluene (10 ml) in the dropping funnel and add this solution within 10 minutes to the phosphorus trichloride solution. Maintain the temperature of the reaction mixture at 20 °C by placing it over an ice bath during the addition. A white precipitate will appear. Stir the solution and reflux it for 2 hrs. Filter the reaction mixture under nitrogen and remove the solvent under vacuum. Filter the white residue using a sintered glass crucible under nitrogen and quickly wash it with dry acetonitrile (30 ml). The washing is to be done very rapidly under vacuum, since the product is partially soluble in acetonitrile. Store the white crystals of *bis* (diisopropylamine)chloro phosphate in a desicator.

Electrochemical synthesis

A low oxidation state of a metal ion can be electrochemically generated. For example, copper(II) can be reduced electrochemically to copper(I) and can be trapped to form an olefin copper(I) complexes. One such example is the reaction of dicyclopentadiene with coper(II) perchlorate hexahydrate under electrolytic condition.

$$\text{dicyclopentadiene} + [Cu(ClO_4)_2]\ 6H_2O \xrightarrow{\ e\ } \text{dicyclopentadiene-Cu-OClO}_3 \qquad .....(3.49)$$

Electrochemical synthesis of dicyclopentadiene copper(I) perchlorate

Take a solution of copper (II) perchlorate hexahydrate (0.74 g, 2 mmol) in acetonitrile (50 ml) in a beaker (100 ml). Add dicyclopentadiene (0.29 g, 2.1 mmol) to this solution. Take two copper plates of 0.1 cm × 2 cm × 6 cm dimensions and place them vertically in parallel position (about 3 cm apart) in the beaker. Connect the copper plates by copper wire fixed by steel clips to a direct current power supply unit. Set the voltage to 10 volt and current 0.1 ampere and pass current through the solution for 6 hrs. Stop passing current if the solution gets heated and on cooling once again pass electricity. One of the electrodes will dissolve in to the solution and a white crystalline compound will be deposited on the other copper plate. Take out the white crystalline precipitates of the complex by decanting the solvent.

Ferrocene derivatives

Ferrocene is an organometallic compound that behaves like aromatic compounds. Different kinds of reactions such as nitration, acylation, arylation, etc can be performed on the aromatic part of the ferrocene. However, in all these reactions there is possibility of oxidation of iron(II) to iron(III); thus, they need to carried out in inert atmosphere. Further to this in some such reactions any oxidized amount of iron obtained are reduced by an external reagent such as stannic chloride or by bisulphites. Nevertheless a reaction condition in which the central metal ion does not get oxidized is to be maintained.

Acetylation of ferrocene is an aromatic substitution reaction on cyclopendienyl ring of ferrocene and is based on the following equation 3.50:

$$.....(3.50)$$

If excess of acetylating agent is used in this reaction, the reaction can lead to diacetylated product. Same reaction can be performed by reacting ferrocene with acetyl chloride in the presence of Lewis acid such as anhydrous aluminium chloride.

The IR spectra of the ferrocene and mono-acetyl ferrocene are given in Fig. 3.42. The characteristic $v_{C=O}$ in mono-acetyl ferrocene is observed at 1656 cm^{-1} which is absent in the ferrocene.

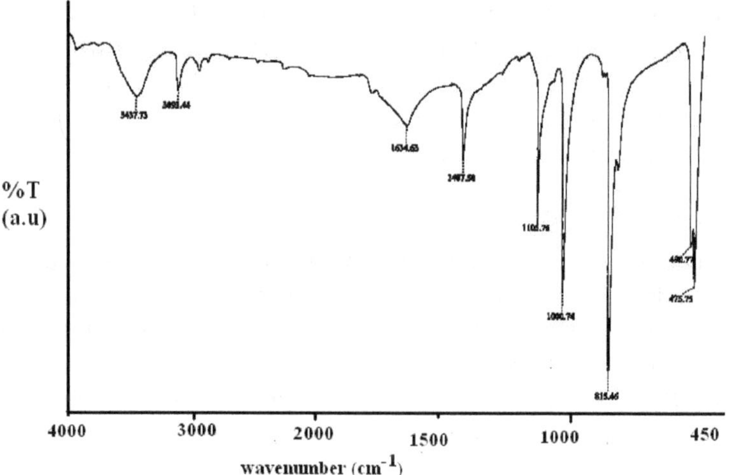

Ferrocene

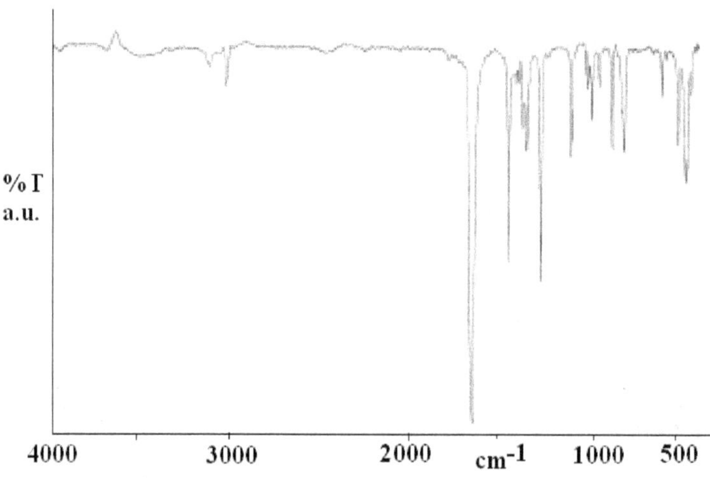

Mono-acetyl ferrocene

Fig. 3.42 FT-IR spectra (KBr, cm^{-1})

Acetylation of ferrocene

Take ferrocene (1.5 g, 8 mmol) and acetic anhydride (5 ml 5.25 g, 87 mmol) in a dry flask (100 ml). To this add phosphoric acid (1 ml) with constant stirring. The solution is protected with guard tube containing anhydrous calcium chloride. Heat the reaction mixture in a steam bath for

15 mins and then pour on to about 20 g of ice in a beaker. Stir the solution and when the ice melts neutralize the reaction mixture by adding solid sodium bicarbonate until carbon dioxide ceases to evolve. Cool the mixture in an ice bath for 30 mins to completely precipitate the acetyl ferrocene from the solution. Filter the solid in a buchner funnel under suction. Wash it with cold water until the filtrate is pale orange. Dry the solid in air for 10 mins. The solid contains ferrocene and mono-acetyl ferrocene. Purify the mixture by column chromatography using silica gel and petroleum ether as eluent. The melting point of mono-acetylferrocene is 80-83 °C.

Heterocyclic ligand

Heterocyclic complexes finds very important place in coordination chemistry. Heterocyclic ligands are either prepared in laboratory or they are commercially obtained. One of the most important ligand used in laboratory is 3,5-dimethylpyrazole. It is easily prepared from the reaction of acetylacetone with hydrazine hydrate.

$$\text{......(3.51)}$$

Preparation of 3,5-dimethylpyrazole

Dissolve hydrazinium sulphate (6.5 g, 0.05 mol) in a sodium hydroxide solution (40 ml of 2.5 M) in a round bottom flask (100 ml). Immerse the flask in an ice bath and when the temperature reaches 15^0C, add acetylacetone (5 g, 5.15 ml, 0.05 mol) drop wise with stirring. Maintain the temperature at 15^0C. When the addition is complete (after about 30 min.), stir for 1 hour at 15^0C; the dimethylpyrazole precipitates during this period. Add water (2 ml) to the reaction mixture and transfer the contents of the flask to separatory funnel and shake with 10 ml of ether. Separate the ether layer; extract the aqueous layers with four 10 ml portions of ether. Wash the combined etheral extract with saturated sodium chloride solution, dry over anhydrous potassium carbonate and remove the ether on a rotary evaporator. The 3,5-dimethylpyrazole will be obtained as white solid.

Tris-(3,5-dimethylpyrazole)copper(II) nitrate

The 3,5-dimethylpyrazole on reaction with copper(II) nitrate forms copper complex *tris*-(3,5-dimethylpyrazole)copper(II) nitrate. The compound is a good catalyst for oxidation reactions and catalyses benzyl alcohol to benzaldehyde in the presence of hydrogen peroxide.The complex has three pyrazole ligands and two nitrate ligands. One nitrate ligand is chelating while the other is mono-dentate. The reaction involved in synthesis and its further reaction and thermal decompositions are represented in scheme 3.11.

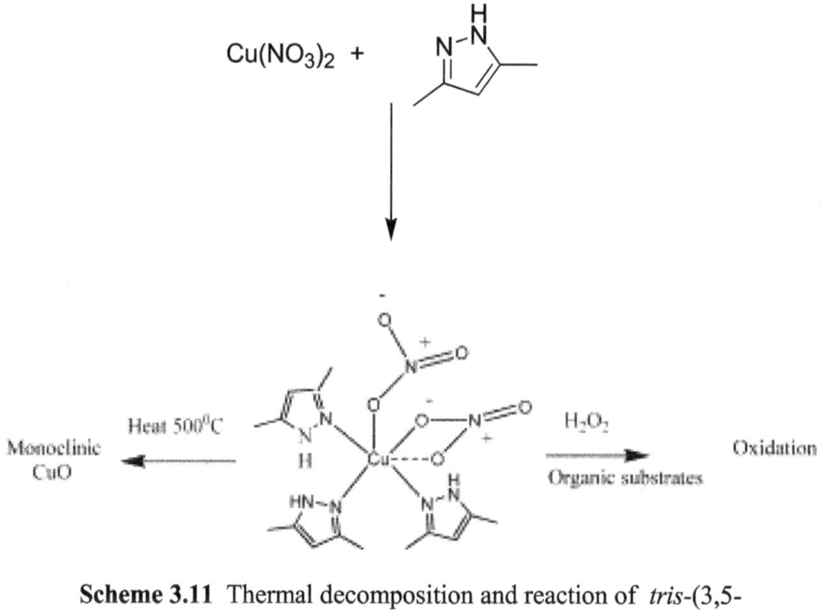

Scheme 3.11 Thermal decomposition and reaction of *tris*-(3,5-dimethylpyrazole)copper(II) nitrate

The IR spectra of the complex has a strong absorption at 1383 cm^{-1} which corresponds to the symmetric 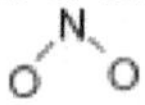 stretching of the NO_3 group and another at 1578 cm^{-1} corresponds to the antisymmetric stretching. A strong absorption at 3242 cm^{-1} due to the N-H stretching frequency of the ligand is present. The complex shows an absorbance at 700 nm ($\varepsilon = 42.1$ M^{-1} cm^{-1}) due to 2E $\rightarrow$ 2T$_2$ transition. The molar conductance of the complex in water is 232 S mol^{-1} cm^2. This molar conductance value corresponds to the complex that ionises in water (equation 3.52).

$$[Cu(dmpyz)_3(NO_3)_2] \rightarrow [Cu(dmpyz)_3]^{2+} + 2NO_3^- \qquad .....(3.52)$$

The magnetic moment of the complex is 1.71 B.M. The value corresponds to spin only value for a single electron arising from d^{-9} electronic configuration. On heating, the complex $[Cu(dmpyz)_3(NO_3)_2]$ complex losses the 3,5-dimethylpyrazole ligand at 140-160 °C. The complex shows a weight loss corresponding to the loss of N_2O_5, to give a residue of CuO in the range of 300-450 °C. Thus, it is seen that decomposition of a complex lead to metal oxides at a relatively low temperature.

So, such decomposition can be useful for making important oxide or related materials. Another example of such decomposition leading to oxide material under mild condition is the decomposition of ammonium dichromate. The decomposition of ammonium dichromate leads to exothermic reaction with formation of nitrogen gas and chromium trioxide. Similarly, alumina fine particles can be prepared by heating aluminium complexes of hydrazide of oxalic acid and particle size in such synthesis can be controlled.

Vitamin B$_{12}$ model Compound

Coenzyme of vitamin B_{12} contains a cobalt-carbon single bond through which adenosyl group is bonded to cobalt. The stability of metal carbon bond in aqueous medium is interesting as metal carbon bonds are generally susceptible to degradation by heat, air, and water. Thus study of such compounds not only allows preparing biologically relevant compounds; but also leaves scope to look for avenues of metal carbon bonded compounds that are stable in water, air etc. Bis(dimethylglyoximato)cobalt complexes, abbreviated as cobaloxamines serve as model for the biological coenzyme. Such cobalt complexes in oxidation states +1, +2, and +3 oxidation states are readily formed making them versatile. Bromo(pyridine)cobaloxamine is an example of such compounds on which wide varieties of reactions can be carried out. The complex is synthesizes on the basis of following equation 3.53.

$$\dots\ (3.53)$$

This compound can be easily converted to carbon-cobalt bond containing complex, namely, methyl(pyridine)cobaloxamine which may be considered as a model compound of vitamin B_{12} (equation 3.54)

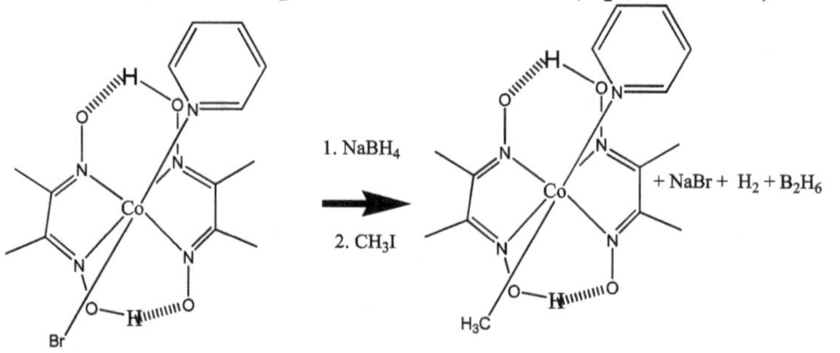

..... (3.54)

Preparation of bromo(pyridine)cobaloxamine

Dissolve dimethylglyoxime (12.5 mmol) in boiling ethanol (30 ml). To this solution add a solution of cobalt (II) nitrate hexahydrate (6.6 mmol) prepared in hot ethanol (30 ml). Boil the reaction mixture for five minutes. Add an aqueous solution of sodium bromide (10 mmol in 10 ml) and boil for further five minutes. To the reaction mixture, add pyridine (5M) solution in tetrahydrofuran (0.75 equivalents with respect to cobalt). After five minutes of boiling the reaction mixture, cool it on an ice bath to 20°C. Purge air through the solution for 20 minutes (this can be done by using a water aspirator). Filter the reaction mixture and rinse the flask with ethanol. Collect the solid crystalline complex. Wash the crystals with portions of water (5 ml) and ethanol (5 ml) followed by diethyl ether (10 ml). Dry the crystals.

Preparation of methyl(pyridine)cobaloxamine

Take bromo(pyridine)cobaloxamine (1.9 mmol) and add methanol (10 ml) in a two necked flask, purge nitrogen gas through the solution. Nitrogen gas is to be purged throughout the reaction to maintain an inert atmosphere. Weigh sodium borohydride (6.6 mmol) and add one half of the hydride to the suspension in small portions with stirring. The solution will have dark blue-black color. Add iodomethane (3.0 mmol) to the reaction mixture and add the remaining portion of sodium borohydride. The solution will be deep red-orange in color. Continue stirring for 15 minutes. If the solution shows blue coloration add an additional drop of the iodomethane. Pour the reaction mixture to ice cold water (20 ml). Collect the solid product by filtration and wash it with water (10 ml) and petroleum ether (20 ml). Recrystallize the product from mixed solvent of dichloromethane and petroleum ether.

CHAPTER 4

Experiments with Organic Compounds

Organic Reactions

General description of Organic reactions

Organic chemistry centers on the chemistry of carbon. Due to the catenation property of carbon, a large numbers of homologues of organic compounds are available. The synthesis and characterization of organic compounds are prime concern of organic chemists. However, for any transformation, the purity of starting compounds and the purity of final products are important. To make any scheme of reactions practically useful, understanding of reaction mechanism is needed. Mechanism of organic reactions is studied by monitoring the course of reactions and by products analysis and also by isolation of intermediates. In the case of organic chemistry, the ease of use of chromatographic techniques makes it relatively easier to study the course of the reactions. Besides conventional methods of estimation of products, the techniques like high performance liquid chromatography, gas chromatography, in situ nuclear magnetic resonance techniques help in ascertaining interconversion of products with time as a reaction proceeds. Moreover, to enhance specificity and selectivity of a reaction a prior strategy is to be developed, unless a serendipitous reaction takes place. For understanding these purposes, several points are to be clarified, they are namely; chemoselectivity, regioselectivity, stereoselectivity and stereospecificity. When a reagent preferentially reacts with one of the two or more functional groups in a chemical reaction then the reaction is called as chemoselective reaction. Highly chemoselective reactions between squarate derivatives and the amino acid cysteine proceed in aqueous solution at neutral pH as shown in equation 4.1. In this reaction only the amino group replaces the alkoxy group, without affecting any other sites in the compound.

Equation 4.1 A chemoselective substitution reaction

Another example of chemoselective reaction is selective deacylation reaction of the sulphonamide derivative as shown in equation 4.2.

Chemoselective reaction

Equation 4.2 Chemoselective deacylation

In this reaction the rate of deacylation from amide group is faster than the sulphonimide group. This relative difference in the rate of the reactions allows competitive deacylation reaction to occur in a selective manner. The term chemospecificity is used for 100% chemoselectivity (equation 4.2). In other words, all chemospecific reactions are chemoselective but not vice versa.

When there is preferential formation of one stereoisomer over another in a chemical reaction, it is called a stereoselective reaction. When the stereoisomers under consideration are enantiomers, the phenomenon is called enantioselectivity. The term enantiomeric excess is used to quantitatively express the process. Similarly, in the case of diastereoisomers, it is called diastereoselective; and when the optical activities of the products are to be quantitatively expressed, the term diastereoisomeric excess is used. Silacyclopropane are useful compounds in stereoselective organic synthesis. The diastereoselective silacyclopropanations of chiral, functionalized alkenes can be done by di-t-butylsilyldichloride with lithium. These reactions lead to highly substituted cyclohexanes having a three member silicon containing ring on it (equation 4.3).

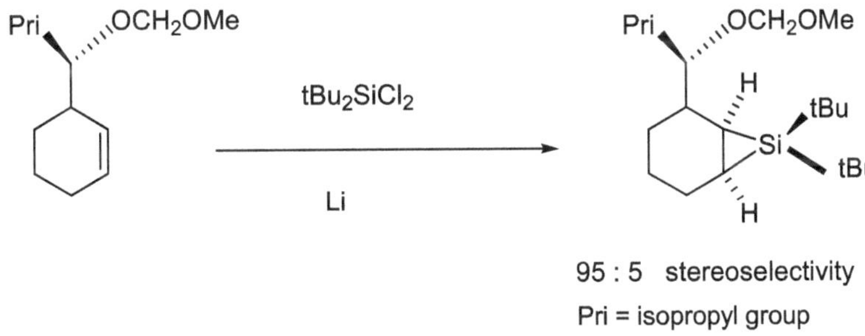

95 : 5   stereoselectivity

Pri = isopropyl group

Equation 4.3 A stereoselective reaction

In a chemical reaction which has potential to lead to the formation of two or more structural isomers, if only one isomer is formed as the sole product; the reaction is called a regioselective reaction. Consider the

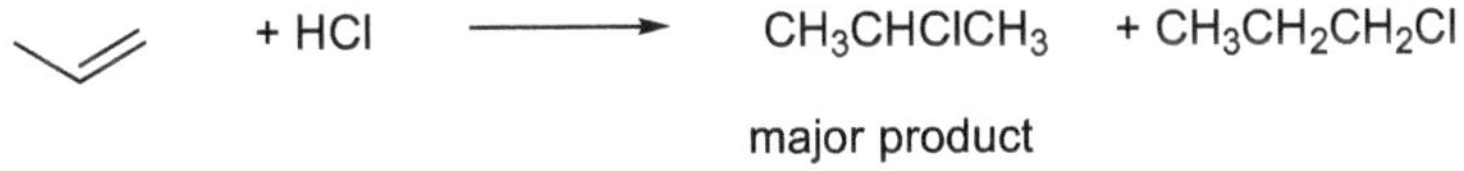

$$+ \ HCl \longrightarrow CH_3CHClCH_3 \ + \ CH_3CH_2CH_2Cl$$

major product

Equation 4.4 Addition of hydrochloric acid to propene

addition reaction of hydrochloric acid to propene. This reaction gives 1-chloropropane and 2-chloropropane as shown in equation 4.4. In this reaction, the 2-chloropropane is formed as the major product; so the reaction is an example of a regioselective reaction.

Regioselective 4+2 cycloaddition reaction among dienes and alkyl substituted vinylboranes takes place on heating. A trivalent boron atom behaves as electron-withdrawing groups. Thus, these groups can activate

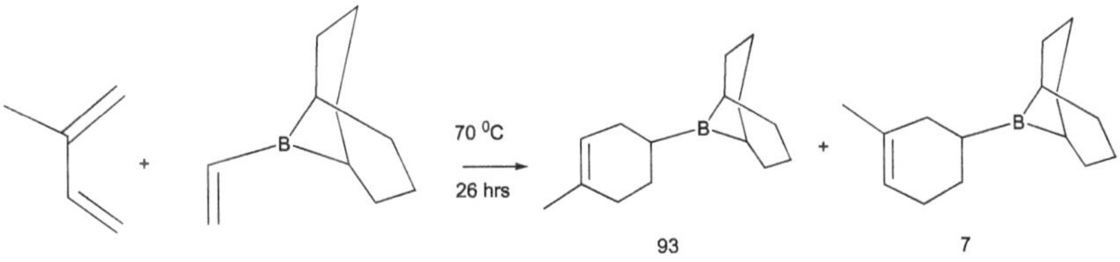

Equation 4.5 An example of 4+2 cycloaddition reaction

dienophile in Diels-Alder (4+2 cycloaddition) reactions. The reactions of 9-vinyl-9-borabicyclo[3.3.1]nonane with various dienes takes place under mild conditions. These reactions take place with high regioselectivity. One illustrative example of such regioselective reaction is shown in equation 4.5

Purity of organic compounds

Purity of chemicals at any stage of chemistry plays an important role in chemistry. Whether it is a chemical reaction or material properties, purity of the chemical compound is the index at which the results can be interpreted. Among the purification techniques sublimation, crystallization, distillation and chromatographic purifications are common techniques and finds important place in day to day chemistry. Apart from these techniques, the sophisticated techniques such as preparative gas chromatography, high performance liquid chromatography, ion-exchange method, etc are also commonly used techniques for purification of chemical compounds. Various types of distillation processes are in use for purification and isolation purpose. Distillation under ordinary condition, distillation under reduced pressure, steam distillation, flash distillation, fractional distillation and steam distillation are commonly used distillation processes. However, each of these distillation methods is used as per the requirement of an experiment. In a simple distillation process the vapor is withdrawn and condensed. This technique is used to separate a volatile liquid from a nonvolatile solute or solid/liquid. The advantage of difference in boiling points of two or more substrates is taken in this process. The fractional distillation is used to separate mixture of liquids that are having very close boiling points, in this process the boiling and condensation process is repeated successively. Steam distillation causes volatilization of a substances by passing steam into a mixture of compound/s suspended in water. The steam distillation is done only when the organic compounds under consideration have appreciable boiling points and different mixing ability with water. Vacuum distillation is carried out under diminished pressure and is used for organic substances that have relatively high boiling points that cannot be distilled satisfactorily under atmospheric pressure, or compounds, which are not stable at high temperature.

Principles of steam distillation

Let us assume that aniline and water are completely immiscible. The vapor pressure of pure aniline are 7.1 kPa at 98 °C and that for water at same temperature is 94.3 kPa. By the law of partial pressure the total vapor pressure of an agitated mixture is the sum of these two vapor pressures; that is 101.4 kPa. A liquid boils when vapor pressure is equal to external pressure. The normal atmospheric pressure is 101.3 kPa under usual condition. This means that aniline water mixture will boil at a temperature near about 98 °C; this boiling temperature is much lower than boiling point of aniline (184 °C). Thus, the normal distillation of these liquids require much high temperature; and boiling temperature of aniline can be brought down by mixing it with water and this boiling temperature is lower than the temperature of steam; so it will be carried by a steam passing through it. Thus, by passing a stream of water vapor can cause distillation of a preheated mixture of aniline water, which is called steam distillation. Such distillation has advantages as some of the compounds decompose on heating to high temperature. Such compounds are possible to distil if it slightly mixes with water and shows above mentioned properties.

Separation by difference in solubility

Based on the difference in solubility of two compounds in different solvents, extraction of the compound by adding appropriate immiscible solvents, the solvent extraction of a compound can be carried out by using separatory funnel. This method is conventionally used for separating an organic compound from inorganic impurity. This is based on the fact that the inorganic compounds generally are ionic are highly soluble in water, whereas organic compounds are soluble in organic solvents. There are many organic solvents that have differences in density with water and are immiscible with water. For this purpose, immiscible solvents are used. For example chloroform and water or hexane and water can be used. In the former case, the lower layer will be chloroform layer as it has higher density than water, whereas in case of water and hexane, the upper layer is hexane layer. The organic acid and bases can be separated by this means, taking advantage of the fact that if an acid or a base is converted to corresponding salts their solubility in water increases. So, by adding water to such sample will make solution in water, the remaining organic compound mixed with organic acid or base will be soluble in organic solvent. Thus they can be extracted. The acid or base, thus taken to an aqueous layer can be easily recovered by neutralization and can either be recovered from aqueous solution by evaporation or by extracting it to an organic solvent and removing the solvent.

Separation of solid mixture of naphthalene and benzoic acid

Take a mixture of naphthalene and benzoic acid (about 2 g) in a beaker (250 ml capacity). Add sodium bicarbonate solution (100 ml, 10%) and stir the suspension for about 15 mins. Recover the insoluble component by filtration. The insoluble compound is naphthalene. The aqueous solution contains sodium benzoate, add dilute hydrochloric acid (10 ml, 20% v/v) to the solution; a white precipitate will appear. Filter the precipitate and dry, this is benzoic acid. Alternatively, naphthalene can be extracted with chloroform from the aqueous solution of sodium benzoate having naphthalene suspended. Dry the chloroform layer over anhydrous sodium sulphate and decant to a round bottom flask. Upon distillation of chloroform gives the naphthalene.

Thin layer chromatography

Small amount of compounds are generally purified by liquid chromatography. This technique is also used to identify number of components in a mixture of compounds, as well as to monitor the progress of a reaction. Thin layer chromatography is one such technique, which is used to either monitor formation of product/s in a chemical reaction or to separate products in a reaction from unreacted reactants or to isolate multiple products formed in a chemical reaction in pure form. Liquid chromatography is based on adsorption phenomenon. In liquid chromatography there are two phases involved in the process of separation. One is a stationary phase and another one is a mobile phase. In thin layer chromatography (TLC), the stationary phase is prepared by mixing silica gel or alumina (with a binder like $CaSO_4$) in a solvent to make slurry. The slurry is applied on a glass or plastic plates to make coating with appropriate thickness. The solvent evaporates and a layer of silica / alumina, gets attached on the plate, it is called the TLC plate. The thin layer on glass plate is called a chromoplate.

With the help of a small capillary tube, the mixture of compounds are placed at one end of a chromoplate. The chromoplate is then placed inside a close chamber containing a solvent with appropriate polarity. The solvent (eluent) moves up the plate through capillary action. Solvent used for elution should have low boiling point. This helps in quick drying of the plate.

The ratio between the distances traveled from the spot by a compound to that of the distance traveled by the solvent is called R_f value of the compound:

$$R_f = \frac{\text{Distance travelled by the compound}}{\text{Distance travelled by the solvent}}$$

The R_f value is a characteristic property of a compound with respect to the solvent under consideration. The R_f values of polar compounds are lesser than non polar compounds. The R_f values have the following trend:

Saturated hydrocarbon < alkenes, alkynes, aromatic hydrocarbons < esters, aldehydes and ketones < amines, alcohols, thiols < phenol < carboxylic acids.

The R_f value increases with increasing polarity of the solvent for a compound. The order of polarity of some common solvents is:

Hexane < cyclohexane < carbontetrachloride < toluene, dichloromethane < chloroform < diethylether < ethylacetate < acetone < propanol < ethanol < methanol.

There are several methods to find out the location of a compound on a TLC plate. For example, observing the spot under UV-lamp, putting the plate in iodine vapor chamber, spraying dilute acidic potassium dichromate or potassium permanganate solution to the plate etc.

Column chromatography is another chromatographic technique. It is based on the process of differential adsorption. The moving phase in this technique is a solvent and the stationary phase is solid adsorbent such as alumina or silica. When solvent containing compounds pass through a stationary phase, adsorption on the surface of the adsorbent separates the components of the mixture. The efficiency of separation of the mixture of compounds by column chromatography depends upon the nature of the: (a) adsorbent; (b) solvent; (c) flow of solvent through the column; and (d) temperature.

Methanol, acetone, ethylacetate, hexane, benzene, diethylether etc are commonly used solvents. When the components of the mixture are colored, the location can be determined visually. For colorless compounds fractions are collected from a column are to be examined by TLC.

Other than these, the commonly used techniques for purification of organic compounds are gas chromatographic and high-pressure liquid chromatographic techniques. In *gas chromatographic technique* the eluent is a gas. The static phase is chromatographic materials that have ability to absorb different molecules at different temperature. A gas chromatograph technique involves use of various detectors such as flame ionization detector, thermal conductivity detector etc. For the former case ionization of materials are carried out by ignition; thus it requires additional gas for combustion over the eluent (nitrogen or helium). The thermal conductivity detector uses a gas with high conductivity and hydrogen finds place for such detection. In *high performance liquid chromatography* the detectors used are UV-visible detector or fluorescence detector etc.

After purification of a sample different techniques are applied to establish their identifications. From chemical analysis point of view the test for elements and functional group identifications are commonly used. Some of the simple tests employed are based on physical properties, such as color, odor, taste, burning properties, melting point, boiling point, sublimation point etc. More importantly various spectroscopic techniques helps in establishing a structure. For this purpose most common spectroscopic techniques are ^1H and $^{13}CNMR$ spectroscopy, 2D-NMR spectroscopy, UV-visible spectroscopy, IR spectroscopy.

Resolution of optical isomers

The solubility differences of various derivatives of optically active compounds are used in optical resolution. They are also separated by high performance liquid chromatography with the aid of chiral columns. However, a racemic mixture of alcohol is generally converted to its racemic hydrogen phthalate ester by heating it with phthalic anhydride and the ester thus formed is resolved by crystallizing with an optically active organic base. The resulting optically active hydrogen phthalate ester is hydrolysed with aqueous sodium hydroxide to regenerate one of the optically active forms of the alcohol. Let us consider the case of resolution of octan-2-ol. The process involved in obtaining chiral alcohol is based on the scheme 4.1.

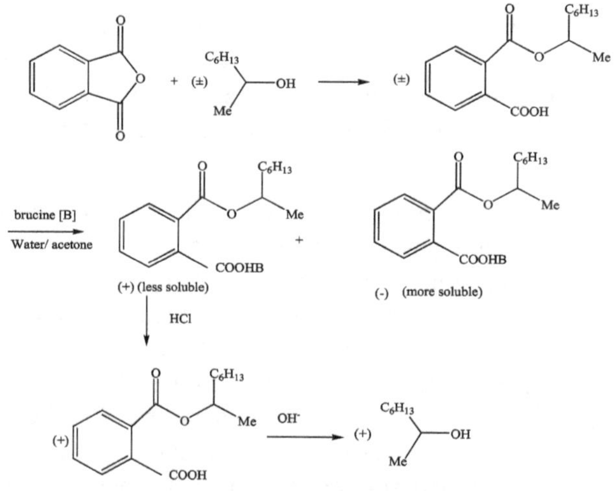

Scheme 4.1 Optical resolution of octan-2-ol

Procedure for resolution

Take dry octane-2-ol (6.5 g, 5 mmol), phthalic anhydride (7.4 g, 5 mmol) and dry pyridine (4 g, 52 mmol) in a flask and heat the reaction mixture for one hr. Cool the reaction mixture and add acetone (5 ml) to it. To this solution add concentrated hydrochloric acid (6 ml) with an equal volume of ice cold water. Add more acetone to make the mixture homogeneous. After this step, add ice-water to the reaction mixture until an oily material completely comes out; this mass sets to a hard solid within 1-2 hrs. Filter the precipitated (±) octyl hydrogenphthalate. Add anhydrous brucine (19.7 g, 5 mmol) to a solution made from finely ground (±)-octyl hydrogen phthalate (1.4 g) in acetone (50 ml). Reflux the mixture to make a clear solution. Cool the solution, the brucine salt crystallises. Filter the crystal and wash them with acetone (10 ml). Combine the filtrate and washing and keep them aside. Cover the crystals of brucine salt with acetone and add dilute hydrochloric acid (6 ml, 1:1). Add ice water to precipitate out the (+) ester completely. Filter the product, wash it with cold water, and dry in air.

Concentrate the filtrate and washing to half of its original volume. Add dilute hydrochloric acid (6 ml, 1:1); then add water (50 ml). Collect the optically active (-)-2-octyl hydrogen phthalate. Crystallise the other form of ester from the remaining solution by adding acetic acid. Add sodium hydroxide solution (1:2 moles ratio to ester) to the respective 2-octyl hydrogen phthalates. Steam distillation of the respective reaction mixtures of the above discussed two independent reactions, optically pure (+) and (-) octan-2-ol will be obtained.

Nuclear magnetic resonance

The ^{1}H or ^{13}C isotopes have I = ½, when these nuclei are placed in a magnetic field, they absorb at a frequency characteristic of the isotopes. A molecule containing hydrogen atoms can show ^{1}HNMR signals as ^{1}H isotope is naturally abundant isotope. However, in the case of carbon ^{12}C is the naturally abundant nuclei; the ^{13}C isotope of carbon is very less abundant nuclei. In order to get good ^{13}CNMR signal it is necessary to accumulate a large number of signals and finally add them up to get a better signal. This is one of the basis of Fourier transform NMR spectroscopy. Alternately, the carbon centers are to be enriched with ^{13}C isotope to get better spectra.

The hydrogen and carbon nuclei in organic compounds are in different local chemical environment, so different protons or carbons in a molecule resonate at different frequencies. Such shifted or unshifted frequencies are directly proportional to the strength of the applied magnetic field, the shifts are converted into a dimensionless quantity known as chemical shift. It is the relative measure from some standard resonance frequency and resonance frequency of a reference. As a reference tetramethylsilane is used for ^1H and ^{13}C nuclei. The frequency shifts are extremely small in comparison to the fundamental NMR frequency. The chemical shift is generally expressed as parts per million or called as ppm.

Consider the ^1HNMR spectrum for ethanol (CH_3CH_2OH). There are three specific signals expected at three specific chemical shifts. The CH_3 group has the highest number of diamagnetic content so it appears at lower ppm value or δ value, relatively CH_2 has less number of hydrogens around carbon and flanked by a carbon and oxygen which make it to appear at higher ppm value than the CH_3 group. The hydrogen on OH is labile; hence it appears at different positions depending on concentration and solvent used. It has a shift around 2-3 ppm depending on the solvent used. Due to unrestricted rotation of the C-H bonds and C-C bonds all the three hydrogens in the methyl group has the same chemical shift. Since the natural abundance of ^1H is very high, intensity of the NMR signals for a set of magnetically equivalent protons will be proportional to the number of protons. So, the integration of a signal and comparison with another signal provides useful information in quantifying the number of equivalent protons.

Hydrogen nuclei are sensitive to the environment. In a general sense this may be connected to hybridisation of the atom to which a set of the protons are attached. Nuclei get deshielded means the effect of electron density gets reduced by electron withdrawing groups. Deshielded nuclei appear at higher ppm or δ values, whereas shielded set of nucleus resonate at lower δ values. The labile protons can be exchanged with other isotopes. For example, the proton of a hydroxy group can be replaced by D_2O to form O-D bond and the particular NMR signal will diminish. Thus, by such an exchange the presence of exchangeable protons can be confirmed.

The nuclei of an element can be considered to be very small magnets when they are associated with magnetic moment. These magnets can interact with each other and this is known as spin-spin coupling. In a compound such interactions between two nuclei occurs through chemical bonds and such effect can prevail upto long range but in NMR it can be typically observed upto three bonds. The magnitude of coupling is expressed in Hertz is called J-coupling.

Coupling to *n* equivalent (spin ½) nuclei splits signal of neighboring signal into *n*+1 multipltes. The intensity ratio for such coupling scheme is shown by Pascal triangle as illustrated in Fig. 4.1.

2	triplet	1 2 1
3	quartet	1 3 3 1
4	pentet	1 4 6 4 1
5	sextet	1 5 10 10 5 1

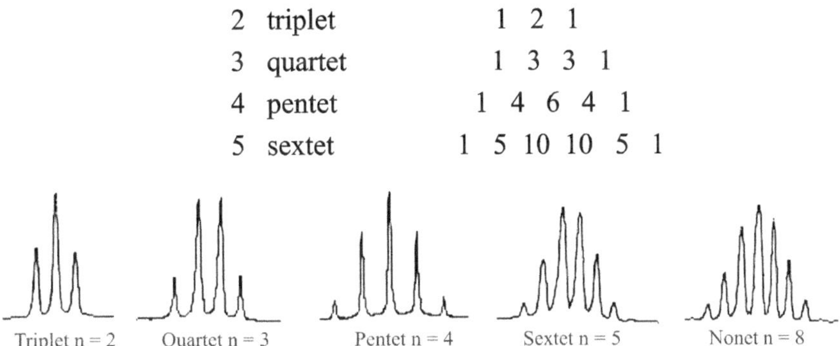

Fig. 4.1 Different splitting patterns of the NMR signals

For example, for CH_3CH_2 group the CH_3 group splits into a *triplet* with intensity ratio of 1:2:1 by the two neighboring CH_2 group. Similarly, the CH_2 gets split into *quartet* with intensity ratio of 1:3:3:1 by the three neighboring CH_3 protons. Coupling to any other nuclei having spin ½ such as phosphorus-31 or fluorine-19 are similar but in such cases the intensities are equal. When one proton interacts with two different protons, a doublet of doublet will be seen (Fig. 4.2). A proton coupling to two other protons of one type, and also to a third proton of another type with two different coupling constants, a triplet of doublets will be observed. A quartet of doublet can arise from simultaneous coupling of a nucleus with three protons in equivalent environment and these further coupling with one more proton at different chemical shift. Correlation spectroscopy (COSY) is a common form of two-dimensional nuclear magnetic resonance (NMR) spectroscopy. In correlation spectra the coupling constants are drawn against the chemical shifts.

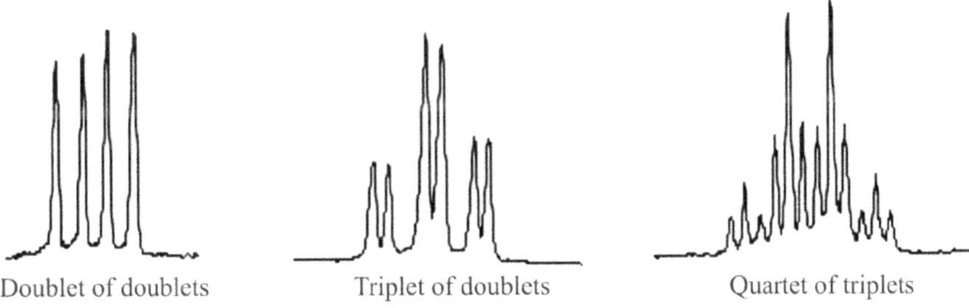

Fig. 4.2 Splitting pattern by set of protons with more nonequivalent protons

Interpretation of spectra

Let us consider the proton NMR spectra (Fig. 4.4) of the sample N-(2,4-dinitro phenyl)-N′-isopropylidene hydrazine which is commonly prepared for identification of acetone by condensation of acetone with 2,4 dinitrophenyl hydrazone. It has different sets of protons that are denoted by a-e in the structure. The two methyl groups are not equivalent as they are *syn* and *anti* to the lone pair of nitrogen and thus they appear as two independent closely spaced singlets at 2.10 and 2.15 ppm. There are three aromatic protons. Ha is flanked by two nitro groups and are not coupled to each other whereas the Hb and Hc are coupled to each other. They appear as two doublet of double with J = 8 Hz, whereas the Ha appears as a singlet. The NH proton appears as singlet at 9 ppm (Fig. 4.4; top).

Fig. 4.3 N-(2,4-dinitro phenyl)-N′- isopropylidene hydrazine

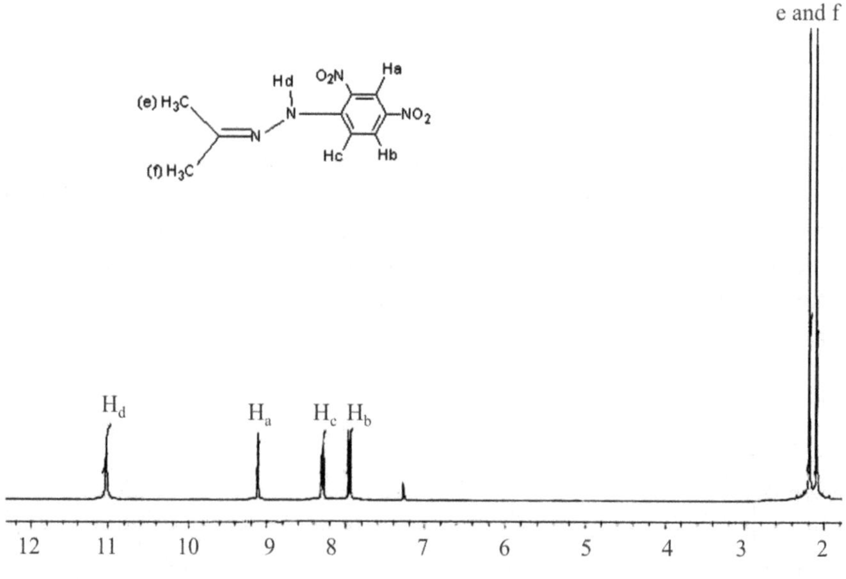

N-(2,4-dinitro phenyl)-N′-isopropylidene hydrazine

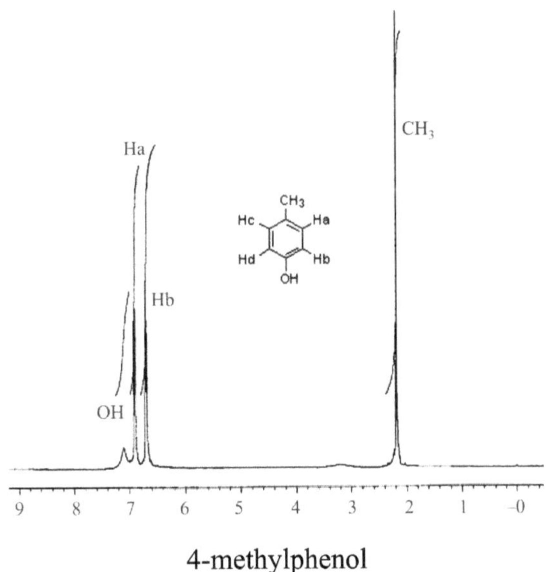

4-methylphenol

Fig. 4.4 ^{1}HNMR spectra (400 MHz, CDCl$_3$, ppm)

The ^{1}HNMR of 4-methylphenol has two aromatic peaks, which appears as two doublets of A$_2$B$_2$ pattern at 6.6 and 6.8 ppm with J = 8 Hz. The methyl signal appears as singlet at 2.2 ppm. The OH signal appears at 7.2 ppm as broad singlet (Fig. 4.4, bottom).

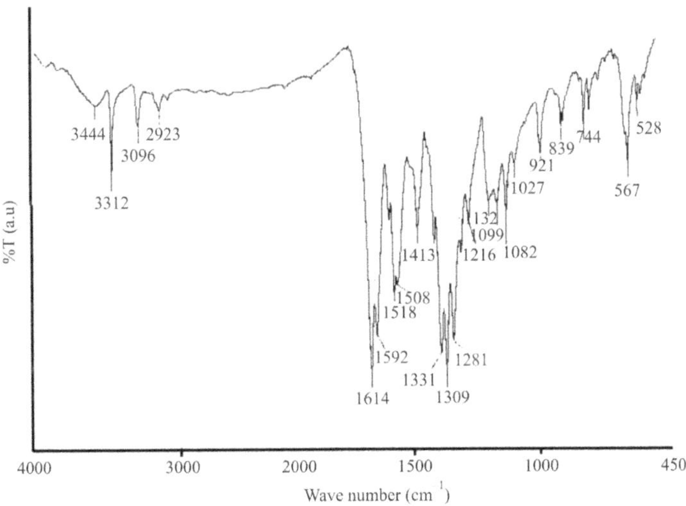

Fig. 4.5 FT-IR spectra (KBr, cm^{-1}) of N-(2,4-dinitro phenyl)-N′- isopropylidene hydrazine

The IR spectra of N-(2,4-dinitro phenyl)-N′-isopropylidene hydrazine characteristic NO_2 frequencies at 1518 cm^{-1} ($v_{asymmetric}$) and 1423 cm^{-1} ($v_{symmetric}$). It has strong peak for $v_{C=C}$ stretching at 1634 cm^{-1} and the absorption for $v_{C=N}$ frequency appears at 1592 cm^{-1}. A broad absorption at 3444 cm^{-1} occurs for the NH stretching frequencies (Fig 4.5).

Mass spectra

Mass spectrometry is used to identify the chemical composition of a compound or mass to charge ratio of different ionised species of a compound. In a mass spectroscopy experiment a sample undergoes chemical fragmentation to form charged particles. The ratio of charge mass of the particles are determined by passing them through electric and magnetic fields. Mass spectra of two commonly used solvents are given in Fig 4.6. In the case of the mass spectra of dichloromethane peaks with m/z at 88, 86 are due to the molecular ion peak of dichloromethane with different isotopic abundances of chlorine. Base peak at m/z 47 corresponds to loss of chloride ion to give CH_2Cl^+ species. Due to the chlorine radical peaks with m/z 36, 37 are observed.

For ethylacetate highest m/z is at 88; corresponds to the molecular ion peak. The base peak for ethylacetate appears at m/z 43 due to the loss OCH_2CH_3 anion to form CH_3CO cationic radical peak. This happens due to stability of acyl cation. The m/z at 29 is due to CH_3CH_2 radical cation.

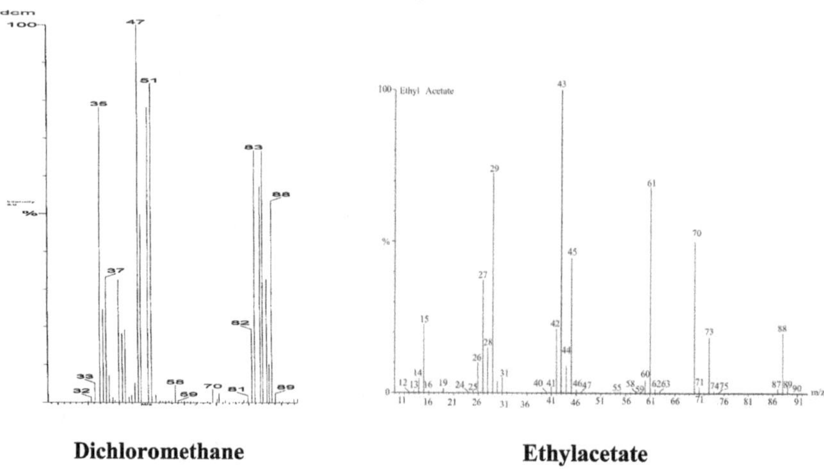

Dichloromethane   **Ethylacetate**

Fig. 4.6 Mass spectra

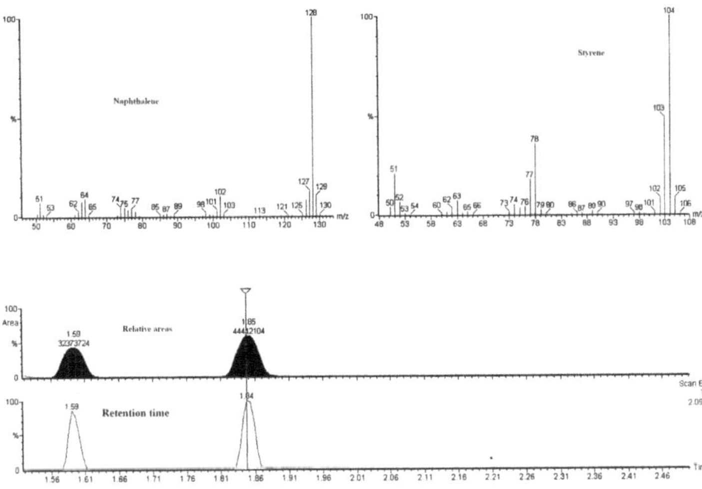

Fig 4.7 GC-MS of a mixture of naphthalene and styrene dissolved in dichloromethane.

Further to this mass spectroscopy coupled with gas chromatography (GC-MS) is used to quantitatively or qualitatively determine multiple numbers of compounds in a reaction mixture. For example, consider the GC- mass spectra of styrene and naphthalene as shown in Fig. 4.7. The top are two mass spectra obtained for naphthalene and styrene; middle one is the relative ratio of the two compounds, the lower one is the retention time for two compounds in gas chromatography. From the GC-MS following can be done: (a) detect how many and what components are present in the mixture. (b) The ratio of the two compounds can be determined from the relative integration area of the peaks. For the later purpose an internal standard is used. The retention time in this case is 1.59 minute is naphthalene whereas the styrene has retention time 1.84 minutes. However, these retention times can vary with the column condition such as flow rate, column materials, type of column used, oven temperature etc. In the mass spectra naphthalene has its m+1 at 129; and the styrene has m+1 at 105.

Naturally occurring organic compounds

Organic chemistry associated with life is indispensable. The plants, animals are made up of carbon compounds. There are many compounds that have some specific activities. Isolation and characterization of biologically active components of natural products is very important not only to use them but also to understand their activity. The active components of biologically important compounds can be directly isolated from its natural source. For example, the different coloring components from vegetables and flowers, starch from potato, sugar from sugar cane, caffeine from tea, lactic acid from milk, ascorbic acid from nutrients etc. can be isolated. In the following part isolation of few naturally available active compounds are illustrated.

Caffeine

Caffeine is a stimulant for central nervous system of human beings. Coffee, tea, soft-drinks contains caffeine. The compound is a white solid and it sublimes at 178 °C. Its structure is shown Fig 4.8.

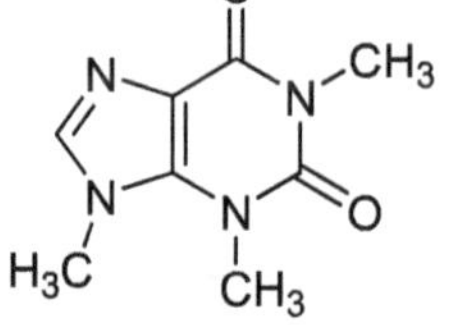

Fig. 4.8 Structure of caffeine

Isolation of caffeine from tea

Add tea (10 g) and powdered calcium carbonate (10 g) to water (100 ml) in a beaker. Boil the contents for about 20 mins. Filter the hot mixture and discard the residue. Cool the filtrate and add dichloromethane (100 ml) to the solution. Separate the organic layer (lower layer) by using a separatory funnel. Dry the dichloromethane over anhydrous sodium sulphate and transfer to a round bottom flask. Evaporate the dichloromethane using rotary evaporator. White solid of caffeine will be left in the flask after removal of solvent.

Lactose

The β-D-galactopyranosyl-1,4-β-D-glucopyranose is known as lactose (also as milk sugar) is a component of milk. Milk contains generally 2–8 % of lactose by weight.

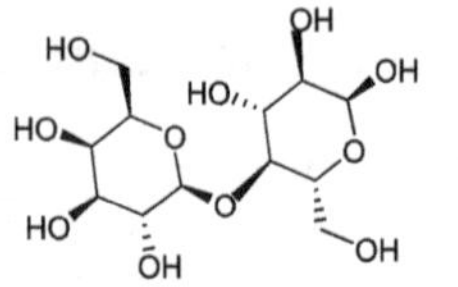

Fig. 4. 9 Structure of lactose

Isolation of lactose from milk powder

Dissolve dry milk powder (10 g) in distilled warm water (50 ml) in a beaker and place it over a water bath at 50 °C. Add acetic acid (10 ml, 10 % v/v) solution and stir the mixture to coagulate. Filter the precipitate (casein) by a passing through a funnel containing cotton; collect the filtrate in a beaker. Add anhydrous calcium carbonate (about 1 gm) to the filtrate. Boil the filtrate with stirring for about 10 mins. Add decolorizing carbon (about 2 g), boil the mixture thoroughly. Filter under suction using a Buchner funnel. Concentrate the filtrate to about 15 ml on a hot plate. Add ethanol (50 ml, 95%) to the concentrated solution. Crystallize lactose by keeping it for 24 hrs. Collect the crystals of lactose by filtration or decantation.

Lycopene

Lycopene is a naturally occurring bright red pigment found in tomato. It has a highly conjugated double bonded structure. Due to high conjugation it shows absorption maxima at 443 nm, 471 nm and 502 nm in hexane with extinction coefficient of the order of 10^5 Lmol^{-1}cm^{-1}.

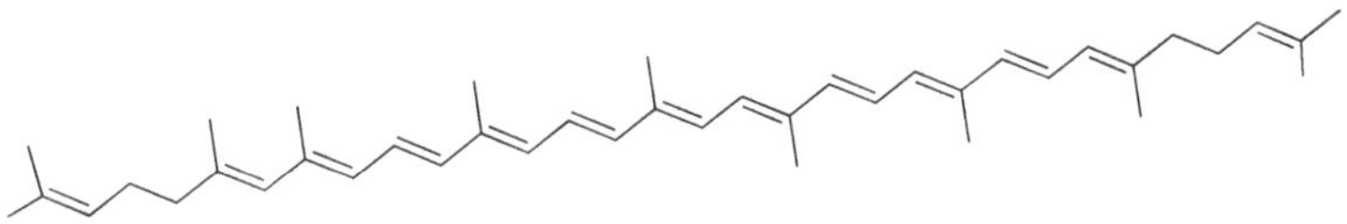

Fig. 4.10 Structure of lycopene

Isolation of lycopene from tomato

Take tomato paste (10 g) in ethanol (25 ml) and dichloromethane (50 ml) in a round bottomed flask. Reflux the solution for about 10 mins. with constant stirring. Filter the solution and discard the insoluble materials. Wash the filtrate with a saturated sodium chloride solution (150 ml). Transfer dichloromethane with the aqueous part to a separatory funnel. The lower layer in dichloromethane contains the crude lycopene. Separate the organic layer and dry over anhydrous sodium sulphate. Evaporate the solvent to obtain crude lycopene

Chlorophyll

Chlorophyll is the green magnesium containing pigment of leaves. It can be extracted by isopropanol from a plant leaves. The chlorophyll is responsible for converting carbondioxide to carbohydrates in nature. It is a macrocyclic type of compound having four pyrole rings connected by intervening carbons. The nitrogen atom of pyrole rings coordinates to magnesium.

Procedure

Take finely torn plant leaves (20 g) in a beaker (500 ml). Add isopropyl alcohol (aprox 250 ml) to soak the leaves. Wrap the beaker with carbon paper. Allow the mixture to stand for overnight. Verify that the isopropyl alcohol solution is green. Filter the solution containing the leaves and evaporate the green filtrate to dryness in a rotary flash evaporator. Transfer the sample to a sample tube after dissolving it in a minimum

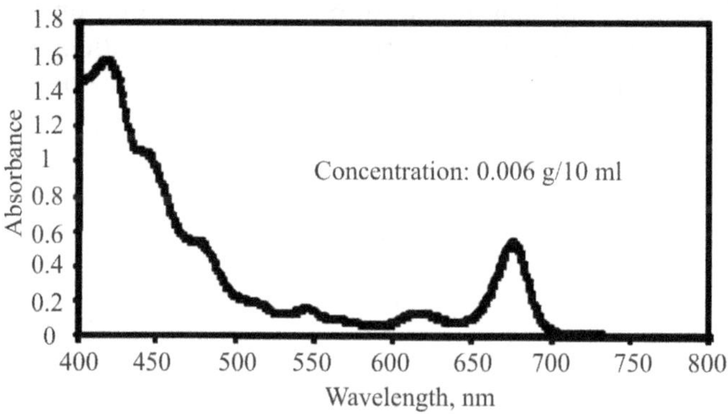

Fig. 4.11 Visible spectra (MeOH) of crude chlorophyll

amount of isopropyl alcohol. Remove the solvent under reduced pressure or on a water bath. Store the crude chlorophyll in dark. A typical visible spectra of crude chlorophyll obtained from leaf is shown in Fig. 4.11. It absorbs around 675 nm showing its characteristic green color.

Piperine from Black pepper

Piperine is naturally occurring compound present in black pepper. It is a conjugate unsaturated amide derivative; having a structure as shown in Fig. 4.12. The compound has also synthetic interest from the point of view that the compound on hydrolysis generates a conjugated dihydroxy aromatic unit attached to a conjugated carboxylic acid containing unit.

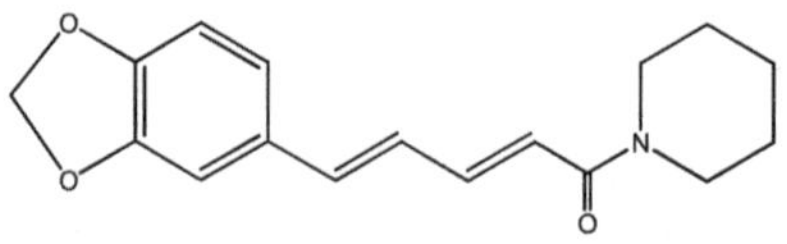

Piperine

Fig. 4.12

Extraction of piperine from black pepper

Add ethanol (about 100 ml) to finely powdered black pepper (10 g) taken in a round bottom flask. Reflux the mixture and distil the ethanol to make it to about 10 ml volume. Add alcoholic potassium hydroxide solution (2N, 15 ml) and warm the solution. Filter the solution and discard the residue. Take the filtrate and concentrate it, obtain a paste, and leave in refrigerator overnight. Solid compound will be obtained. Recrystallise the solid from acetone a yellow solid of piperine will be obtained (m.p. 129-130 °C)

Ascorbic acid in food

Sometimes it is essential to quantify the amount of active components in organic substances. For example, food contains vitamin C which is in the form of ascorbic acid. The amount of ascorbic acid in a sample can be determined by oxidising ascorbic acid with 2,6-dichloroindophenol. The reaction involved in the estimation procedure is shown in equation 4.6.

Ascorbic acid

$$.....(4.6)$$

Estimation of ascorbic acid

Take rice powder (10 g) in a beaker and add phosphoric acid (70 ml) to it. Stir the mixture for about five minutes. Filter the extract through cheese cloth. Discard the insoluble solid and make the volume to 100 ml by adding water. Take 10 ml of the sample extract, place in a conical flask and titrate it with a 2,6-dichloroindophenol solution and find out the end point from the color change.

Organic reactions

On the basis of the type of reactivity such as substitution, addition, elimination, rearrangement, oxidation, reduction reactions, the organic reactions can be divided in to many categories. Each of these reactions are governed by their own mechanistic schemes. Under each category large number of chemical reactions can be listed. For example, hydrogenation, halogenations, halo-hydrogenation, dihydroxylation, hydroboration, hydrosilylation are commonly encountered organic addition reactions.

Hydrogenation

Hydrogenation reaction is considered to be a simple reaction. It involves addition of hydrogen, in the form of hydride or dihydrogen. Generally, catalysts are needed for such reactions of hydrogen gas. The hydrogenation reactions are industrially applied to process vegetable oils and fats. Important physical properties of fats and oils such as the melting point, depends on the degree of saturation. Platinum group metals are used as effective catalysts for hydrogenation reactions. Nickel catalysts in the form called, Raney nickel is commonly used as hydrogenating catalyst. Raney nickel comprises of nickel-aluminium alloy and is used in many industrial processes. Raney nickel is produced by treating nickel-aluminium alloy with concentrated sodium hydroxide. A typical catalyst contains about 85-percent nickel by mass, having overall 2:1 ratio of nickel to aluminium. It causes hydrogenation as well as desulphurization of thiophene derivative as illustrated in equation 4.7.

$$\dots\dots(4.7)$$

Another catalyst commonly used for hydrogenation of organic functional groups is Lindlard catalyst. It consists of palladium deposited on calcium carbonate and treated with different lead complexes. The lead additive passivates the palladium sites. Another very good hydrogenating catalyst is Adam's catalyst. Adams' catalyst is platinum(IV) oxide hydrate. The platinum oxide is not an active catalyst, but it becomes active on reaction with hydrogen which converts it to platinum metal. Adams' catalyst is prepared from ammonium hexachloroplatinate(IV), or from platinic acid by fusing it with sodium nitrate:

$$H_2PtCl_6 + 6\ NaNO_3 \rightarrow Pt(NO_3)_4 + 6\ NaCl\ _{(aq)} + HNO_3$$

$$\dots\dots(4.8)$$

$$Pt(NO_3)_4 \rightarrow PtO_2 + 4\ NO_2 + O_2$$

$$\dots\dots(4.9)$$

The resulting solid from these reactions is washed with water to remove nitrate's impurities. It can be stored in a desicator for further use in hydrogenation reactions.

Selectivity in hydrogenation reactions are always sought after. One selective catalyst is the iridium complex of cyclooctadiene, tricyclohexyl phosphine and pyridine. The catalyst is very selective and has very high efficiency. One highly selective reduction by using this catalyst is shown in equation 4.10.

$$.....(4.10)$$

Another example of selective hydrogenation is the selective reduction of dicyclopentadiene by nickel catalyst as shown in equation 4.11. In this reaction one of the double bond can be saturated selectively and only endo compound is formed.

$$.....(4.11)$$

Selective hydrogenation of dicyclopentadiene to 3,6-dihydro-endo dicylopentadiene

Take nickel(II) acetate tetrahydrate (3.75g, 15 mmol) in a round bottom flask and dissolve it in ethanol (20 ml). Flush the reaction vessel with hydrogen gas and add sodium borohydride solution (15 ml, 0.1M in ethanol). This will reduce nickel acetate and make hydrogenation catalyst. Add dicyclopentadiene (40.7 g, 0.3 mol) in ethanol (200 ml) to the above solution. Inject the slurry thus prepared to a hydrogenating apparatus with stirring. When hydrogen uptake ceases, add activated carbon (1 g) to remove the catalyst. Filter the solution and wash with acetone two times (10 ml each). Combined organic layer is distilled to obtain the hydrogenated product as an oil.

Bromination

Unsaturated compounds can be brominated by reacting with bromine. This can also be done by brominating solutions prepared by reacting potassium bromate with potassium bromide in acidic condition.

$$KBrO_3 + KBr + HCl \rightarrow Br_2 + KCl + H_2O \qquad .....(4.12)$$

This oxidation reaction is also useful for test of unsaturation, to estimate the unsaturation of an olefinic compound. The bromine liberated from such reaction reacts with unsaturated hydrocarbons and the excess bromine liberated from a known amount of bromate can be determined by a volumetric titration.

Let us consider the reaction of chalcone with bromine. In this reaction the chalcone is dibrominated. This reaction is done by taking a solution of chalcone in ethanol under ice-cooled condition in a round bottomed flask.

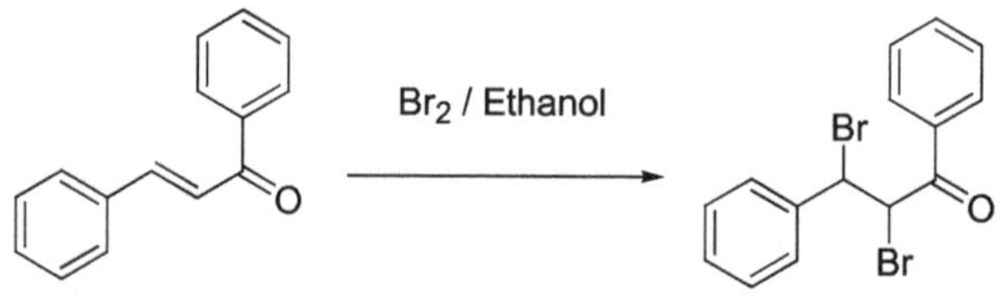

$$.....(4.13)$$

Dropwise addition of bromine to this solution by a pressure equalizing funnel leads to a white precipitate. The amount of bromine consumed can be determined by titrating excess bromine and subtracting this from the known concentration of added bromine. Thus, such reactions are used for test of unsaturation as well as to find out of extent of unsaturation. However, some bromination reaction gets complicated. For example the reaction of some chalcone derivatives lead to incorporation of alkoxy group as well as bromine group as illustrated in scheme 4.2. This happens due to the participation of lone pair of methoxy group in resonance. The methoxy group makes the benzylic bromine more reactive to form a carbocation at that place (scheme 4.2).

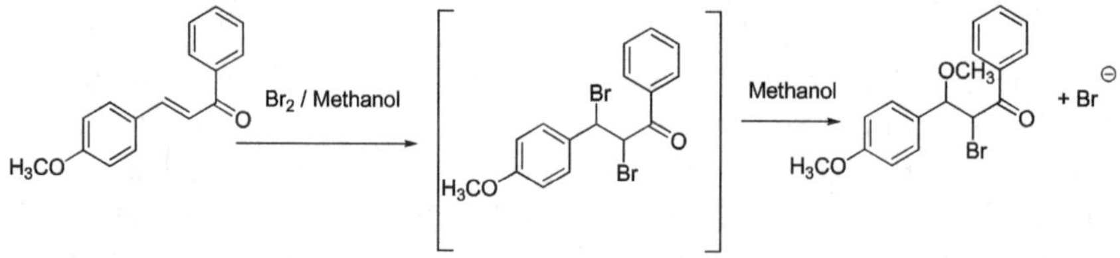

Scheme 4.2 Introduction of methoxy and bromo group to a chalcone derivative

Electrophilic substitution by bromine

Bromine is extensively used for electrophilc substitution reactions of an aromatic compound. The bromination of aromatic is fast and upon bromination aromatic ring gets further activated, so the dibromination and even tribomination takes place in aromatic compounds. An example for selective bromination of phenol by using molecular bromine is shown in scheme 4.3.

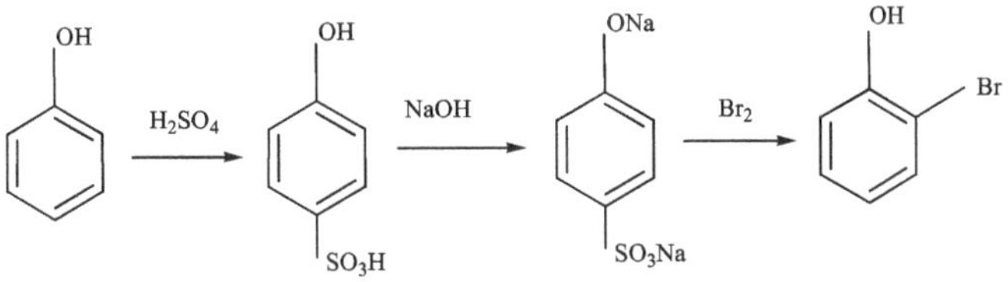

Scheme 4.3 Selective bromination on phenol

This reaction can lead to the tribrominated product. Hence to prepare ortho-brominated phenol; the phenol is first sulphonated at para-position and sulphonated derivative is brominated with bromine at ortho-position and the sulphonate group is oxidized to give the desired ortho-bromo phenol.

Procedure

Place a mixture of phenol (3.1 g, 0.33 mol) and concentrated sulphuric acid (11.6 ml, 32N), in a three necked round bottom flask. Heat the solution on a boiling water bath for 3 hours with mechanical stirring. Cool it to room temperature or below by immersing the flask in ice water, and then slowly add a solution of sodium hydroxide (9.5 g in 250 ml water) to the reaction mixture; solid salt may separate on cooling, but this will dissolve at a later stage. Place a thermometer, which dips well into the liquid, and place a small dropping funnel over the top of the condenser in one mouth of round bottom flask. Cool the alkaline solution to room temperature and then add bromine (17 ml) from the dropping funnel down the condenser during 20–30 minutes. Stir the reaction mixture and control the temperature in the range of 40-50 °C. Continue stirring for 30 minutes. Steam distil the solution to obtain *ortho*-bromophenol.

Nucleophilic substitution reactions

A simple example of nucleophilic substitution reaction is conversion of alkyl bromide into an alcohol (equation 4.14).

$$CH_3Br + OH^- \longrightarrow CH_3OH + Br^-$$

.....(4.14)

Depending on the order of nucleophilic substitution reactions, they are called unimolecular $S_N{}^1$ or bimolecular $S_N{}^2$ type of substitution reactions. Preparation of various ethers is done by reactions of aliphatic halides with alcohols or phenols with potassium carbonate in acetone. Among the various alkyl or aryl halides, allylic halides are more reactive and lead to ether formation under relatively mild conditions. For example, the reaction of 4-hydroxy benzaldehyde with allyl bromide in the presence of anhydrous potassium carbonate and acetone leads to the corresponding O-allylated ether as shown in equation 4.15.

$$.....(4.15)$$

Procedure

Take a mixture of *p*-hydroxy benzaldehyde (4.5 g, 3.7 mmol), potassium carbonate (5.2 g, 3.7 mmol) in a round bottom flask. Add acetone (40 ml) to the solution and stir the solution. While stirring the solution, add allyl bromide (3.2 ml, 3.7 mmol) to the solution in the flask, and stir for another 9 hrs. Filter the reaction mixture. Wash the residue with acetone. Collect the filtrate and concentrate. Extract the organic part by ethylacetate with the aid of a separatory funnel. Concentrate the ethylacetate layer and purify the product by column chromatography.

Debromination reactions

The reactions of 1,2-dibromo alkanes or bromo alkanes with alcoholic potassium hydroxide leads to olefinic compounds by elimination reaction. For example, the reaction of dibromoethane with potassium hydroxide in ethanol gives vinyl bromide. The vinyl bromide is a low boiling liquid (b.p. 16 °C at 750 mm); thus being volatile, it needs extra care for storage. For dehydrobromination reactions, in which volatile olefins are produced, the olefinic compounds are generated by allowing to mixture of dibromo compounds to react with hot alcoholic solution of potassium hydroxide in closed vessel. From such vessel the olefins are directly trapped in gas traps immersed in liquid nitrogen or ice. For example, the vinyl bromide is directly generated and collected in a container kept at low temperature.

$$.....(4.16)$$

Vinyl bromide is a very good precursor for polymerization, Diels-Alder reactions, and orgnanometallic reactions leading to various organic compounds. Vinyl bromide on heating with cyclopentadiene exclusively gives endo-norbonene bromide. This compound is used as starting material for preparation of various norbonene derivatives. For this reaction the cyclopentadiene is freshly produced by cracking; i.e. heating dicyclopentadiene at high temperature over 250 °C (equation 4.17).

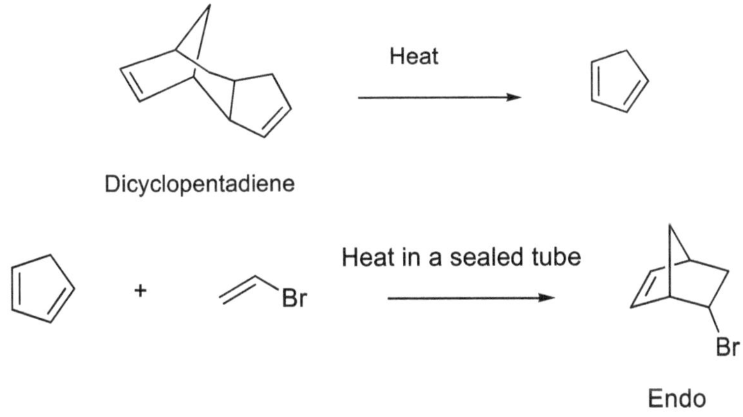

Dicyclopentadiene

Endo

.....(4.17)

Synthesis of *endo*-dehydrnorbonylbromide

Take cyclopentadiene (9.2 g, 1.38 mol) and vinyl bromide (1.69 g, 1.58 mole) in a closed tube connected with a teflon stopper. Close the outlet and place the sealed tube on an oil bath at 160 °C for 12 hrs. Transfer the reaction mixture after cooling. Care should be taken so that the stopper is not opened without cooling the reaction mixture. As careless operation with the inside pressure may lead to explosion. Distil the solution to remove unreacted starting materials.

Diels-Alder reaction

Diels-Alder reaction is one of the most important reaction for the synthesis of alicyclic compounds. It is a [4+2]-cycloaddition reaction in which a conjugated diene and a dienophile (alkene or alkyne) gives the addition product. It involves the interaction of four π-electrons of the diene and two π-electrons of the dienophile. It is a thermally allowed process and the overlap of the molecular orbitals (MOs) is shown in Fig. 4.13.

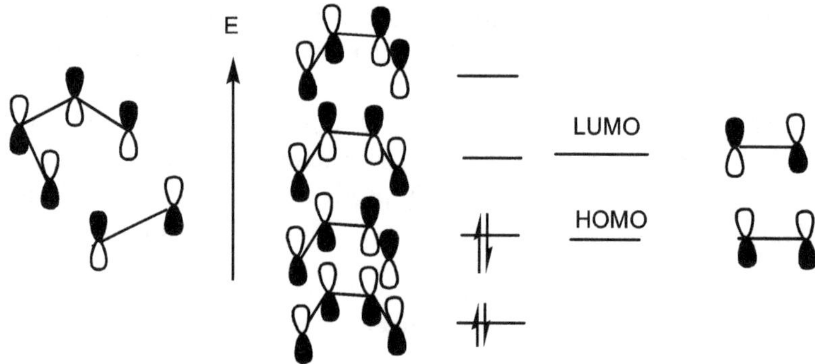

Fig. 4.13 Orbital overlap pictures of different molecular orbitals of butadiene and ethylene

When the orbitals are of similar energy, overlap takes place between the highest occupied molecular orbital (Fig. 4.13) of the diene (HOMO) and the lowest unoccupied molecular orbital of the dienophile (LUMO). The reaction is facilitated by electron-withdrawing groups on the dienophile, since this lowers the energy of the LUMO. The reaction proceeds faster when the dienophiles bears substituents like CHO, COR, COOR, CN, C=C, Ph, or halogen etc. The diene component should be electron-rich as it raises the reaction rate. For example, methyl derivatives of 1, 3-butadiene is more reactive than its unsubstituted diene.

Example of a Diels-Alder Reaction

The reaction between 1, 3-butadiene and maleic anhydride (dienophile) produces 4-cyclohexene-cis-1,2-dicarboxylic anhydride. The 1,3-butadiene is prepared by heating butadiene sulphone to approximately 200°C, which generates sulphur dioxide as a by product. The product is precipitated from the reaction mixture using toluene-hexane mixed solvents. Since the reaction takes place at a high temperature, excess xylene is required to maintain constant temperature. The synthetic path is shown in scheme 4.4.

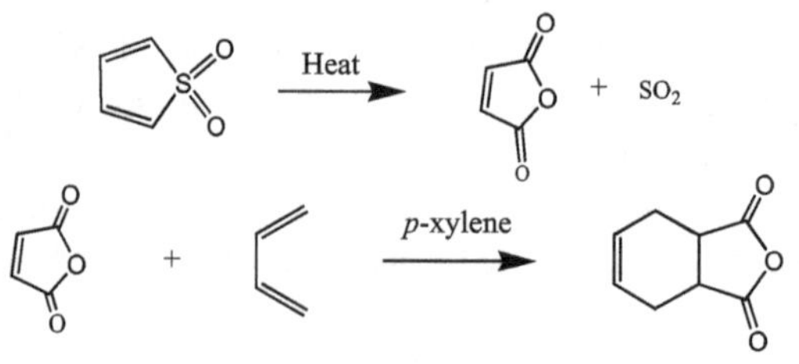

Scheme 4.4

Procedure

Take butadiene sulfone (5 g, 0.05 mol) or maleic anhydride (5 g, 0.05 mol) with xylene (10 ml) in a round bottom flask attached to a reflux condenser on an oil bath. Reflux the reaction mixture for 45 minutes. Cool the reaction mixture to room temperature and add toluene (50 ml). Warm the reaction mixture by adding activated charcoal and filter it. Add hexane (25 ml) to the solution and concentrate the solution by heating at 80 °C. Cool the solution in an ice bath, crystallization will take place. Collect the crystals by suction.

Chlordane

Chlorodane is a commercially available insecticide. It is prepared by 4+2 cycloaddition reaction between hexachloro cyclopentadiene and cyclopentadiene. The hexachloro cyclopentadiene acts as dieneophile whereas cyclopentadiene act as the diene. The cyclo added product, on further reaction with chlorine gives chlordane.

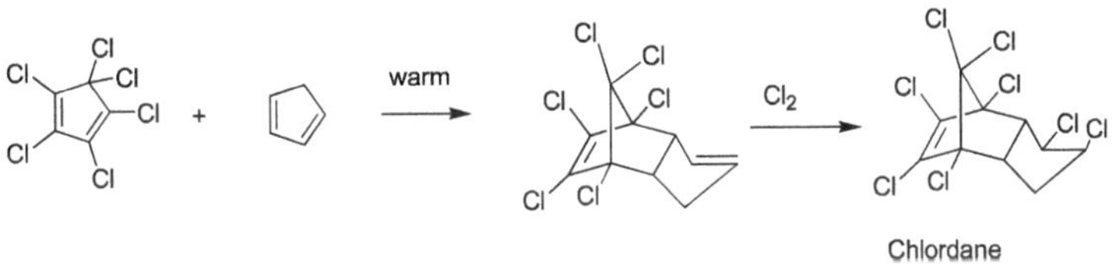

Scheme 4.5 Synthesis of chlorodane

Bromination by radical

Aliphatic hydrocarbons can be halogenated by exposing solutions of hydrocarbons and halogens to ultra-violet light. But the bromo-hydrocarbons can also prepared by reaction of N-bromo succinamide with benzoyl peroxide in the presence of ordinary light. In this regard allylic bromination reactions are conventional examples of such reactions. For example, cyclohexenyl bromide is prepared by reaction of cyclohexene with N-bromosuccinamide in the presence of benzoyl peroxide (equation 4.18). The role of dibenzoyl peroxide is to initiate the formation of bromide radical.

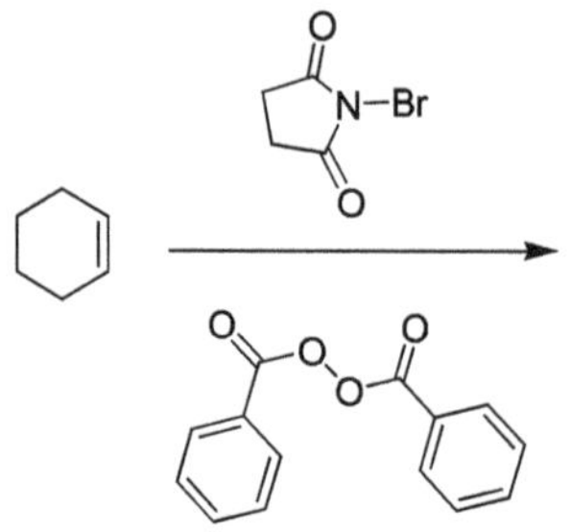

.....(4.18)

Procedure

Take N-bromosuccinamide (3.94 g, 0.20 mol), cyclohexene (4.92 g, 0.60 mol) with carbon tetrachloride (50 ml) in a round bottom flask. Add about 100 mg of benzoyl peroxide to the solution. Reflux the solution, so as to start the reaction. A white suspension of succinamide will appear in the solution. When the reaction slowed down, reflux it for about 1hr. Cool the flask and filter. Wash the precipitate with small amount of carbon tetrachloride and collect the filtrate. Remove the unreacted cyclohexene (b.p. 63 °C) and carbon tetrachloride (b.p.77 °C). Distil the residual liquid product under reduced pressure to obtain 3-bromo-cyclohexene as colorless liquid (b.p. 66-67 °C at 20 mm of Hg).

Nucleophilic addition cum elimination reactions

The chemistry of carbon anion in organic chemistry is vast. It is used for synthesis of various carbon-carbon bond formations and is widely studied. The condensation of a ketone having α-hydrogen to another carbonyl compound leads to hydroxyl ketones, which easily gets dehydrated to give a,β-unsaturated carbonyl compounds. One such reaction is synthesis of dibenzylidine acetone. It is prepared by condensing two equivalents of benzaldehyde with one equivalent of acetone.

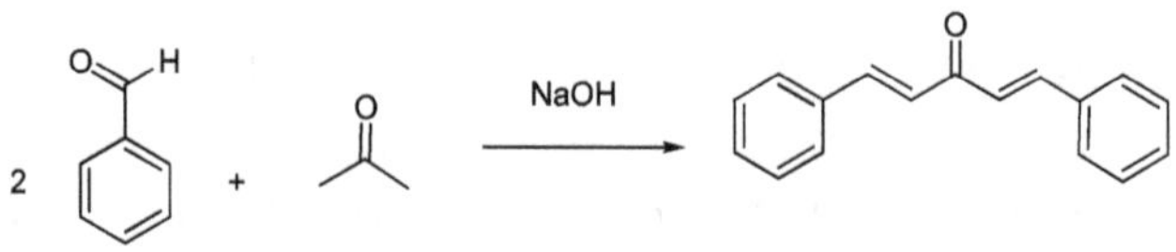

.....(4.19)

The reaction involves generation of an anion of acetone followed by reaction with benzaldehyde and elimination of water from the hydroxyl ketone formed as intermediate in the reaction. The a,β- unsaturated carbonyl compound thus formed undergoes further condensation with another benzaldehyde molecule to give the product (scheme 4.6).

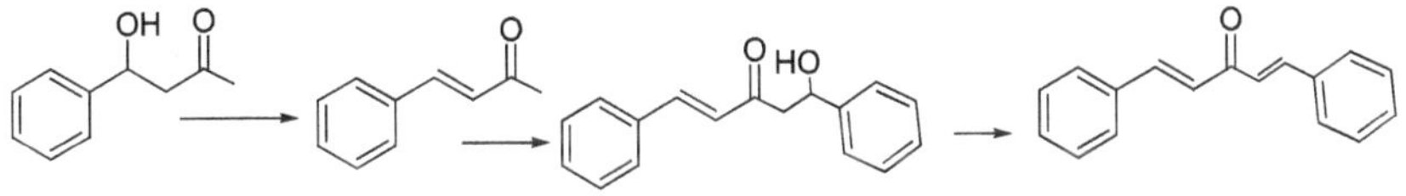

Scheme 4. 6 The steps in synthesis of dibenzylidine actone

Procedure

To a well stirred solution sodium hydroxide (2.48 g in 25 ml of water and 20 ml of ethanol) in a round bottom flask placed on a water bath at 20 °C, add half of the amount of a solution containing benzaldehyde (2.55 ml, 0.25 mol) and acetone (9 ml, 0.13 mol) to the reaction mixture while stirring. A flocculent precipitate will be formed in 2-3 minutes. Add the remaining benzaldehyde-acetone mixture to the reaction mixture after 15 minutes. Continue stirring for 30 minutes. Filter the residue and wash the residue with cold water to remove sodium hydroxide. Dry the solid at room temperature. Recrystallise the solid from hot ethanol to obtain pure dibenzyllidene acetone (m. p. 122 °C).

Perkin Reaction

Perkin reaction is an example of the reaction of carbon anion one example is the reaction between benzaldehyde with acetic anhydride in the presence of potassium acetate.

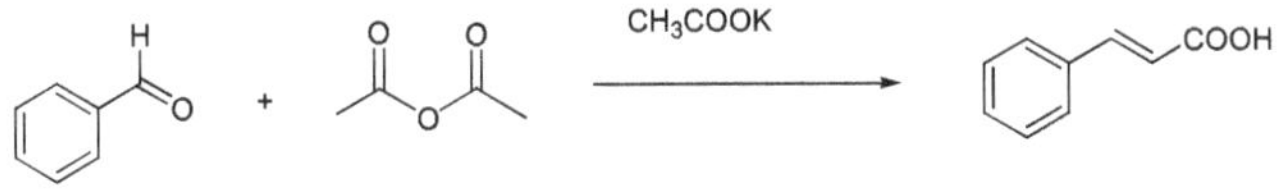

$$.....(4.20)$$

This reaction passes through condensation, elimination and deacylation steps. The later two steps are as shown in scheme 4.6.

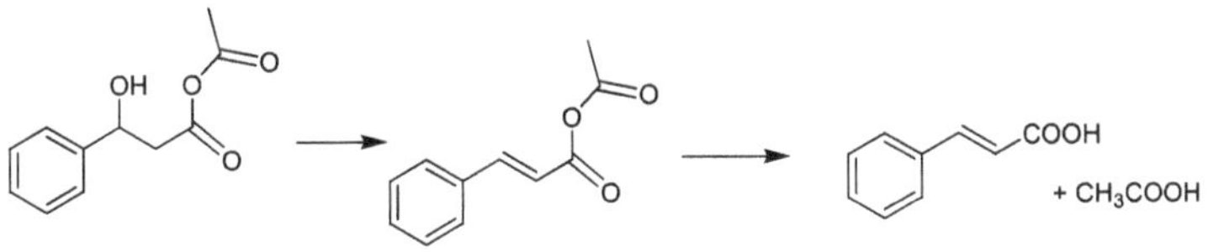

Scheme 4.6 Elimination and deacylation steps in cinnamic acid synthesis

Procedure for preparation of cinnamic acid

Take benzaldehyde (2.1 g, 0.20 mol), acetic anhydride (2.8 ml, 0.29 mol) and freshly fused finely powdered potassium acetate (1.2 g, 0.012 mol) in a dry flask, fitted with a condenser with calcium chloride guard tube at upper end. Heat the reaction mixture on an oil bath at 160 °C for one hour and then at 170-180 °C for 3 hrs. Cool the reaction mixture and add water (10 ml) to the reaction mixture. Add a saturated solution of sodium carbonate and make the medium alkaline. Extract the organic part with diethyl ether (50 ml) using a separatory funnel. Cool the aqueous layer in an ice bath and acidify it with concentrated hydrochloric acid, add acid until the evolution of carbon dioxide ceases. Cinnamic acid will precipitate as white solid, filter the white solid and wash it with cold water and dry it.

A multistep one pot reaction

There are certain reactions that can be carried out in one pot by carefully carrying out different steps by adding reagent in sequential manner. The prerequisite for such reactions is that they should occur in near quantitative yield and in selective manner in the intermediate steps. Dimedone is an active methylene compound, can be synthesized via such reaction schemes in one pot. In this reaction mesityl oxide is reacted with diethylmalonate to obtain a C-C bonded compound through Michael addition reaction. In this reaction, sodium ethoxide is used as base to generate anion of diethylmalonate, which reacts with α,β-unsaturated carbonyl compound mesityl oxide at the γ-position. The product thus formed in this step undergoes cyclisation reaction to form a β-ketoester. β-ketoester gets hydrolysed and easily decarboxylate to the dimedone. Thus, the reaction involves three steps, one is conjugate addition, second is cyclisation, third is ester hydrolysis, fourth step is the decarboxylation reaction (Scheme 4.7).

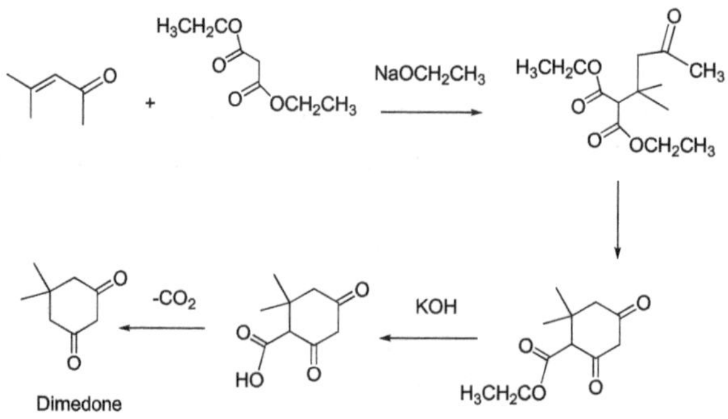

Scheme 4.7

Synthesis of Dimedone

Take clean cut pieces of sodium (2.3 g, 0.1 mol) in dry ethanol (40 ml) in a three necked round bottomed flask. Connect two necks to two dropping funnels one containing diethylmalonate and other containing mesityloxide. The third outlet should be connected to a condenser. Stir the sodium suspended in ethanol so that all the sodium dissolves. Care should be taken while performing this as the exothermic reaction with evolution of hydrogen takes place. Avoid contact with water or any contaminated water in any of the glasswares or reagents. Add dropwise diethylmalonate (16.1 ml, 0.1 mol) to the sodium ethoxide solution and from the other dropping funnel add mesityloxide (11.6 ml, 0.1 mol) dropwise to the reaction mixture. Remove the dropping funnels and cork the two ends and cork the two outlets. Reflux the reaction mixture for two hours and add potassium hydroxide (12.5 g in 60 ml water) after cooling the reaction mixture. Reflux the solution for two hours. Acidify the reaction mixture with hydrochloric acid (1:2 v/v in water). Distil the alcohol as much as possible. Acidify the residue further with hydrochloric acid until it becomes acidic. Cool the acidic solution in refrigerator. Crystals of dimedone will be formed, filter them and characterize the product by recording m.p. (m.p.147-148 °C) and by recording IR and NMR spectra.

Multicomponent reactions (MCRs)

Multicomponent reactions are reactions involving three or more starting reactants to form one or more products in which all or most of the reactants contribute to the newly formed product/s. Generally, several reaction equilibria are combined together resulting in an irreversible step leading to the formation of product. The important point in MCR is to push the multiple equilibria to form one equilibrium yielding product. Thus, the success of such reactions rely on the reaction conditions: solvent, temperature, catalyst, concentration, the nature of starting materials and functional groups attached to a substrate. There are a large number of examples of multi component reaction. One example of such reaction is the reaction between an aromatic azide and ammonium chloride and phenylacetylene. This multicomponent reaction is catalysed by cuprous iodide is shown in equation 4.21.

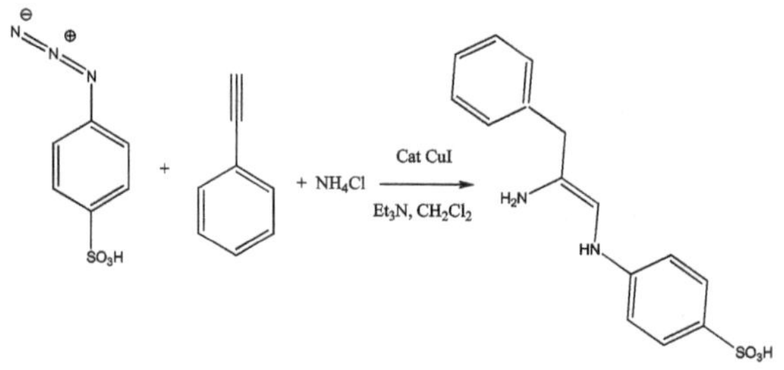

.....(4.21)

Procedure: Take a mixture of ammonium chloride (26.7 mg, 0.5 mmol), cuprous iodide (9.5 mg, 0.05 mmol), and phenylacetylene (51.1 mg, 0.5 mmol) in dichloromethane (1 ml). Add dropwise triethylamine (0.14 ml, 0.75 mmol) to the mixture, and stit the reaction mixture at room temperature. Once the color of the solution changes to yellow add *p*-toluenesulfonyl azide (98.6 mg, 0.5 mmol) dropwise to the reaction mixture. After 12 hrs of stirring add dichloromethane (2 ml) and aqueous ammonium chloride solution (3 ml) and stir for half an hour. Separate the aqueous layer from dichloromethane layer. Lower layer is dichloromethane; further extract the aqueous layer with dichloromethane (3×3 ml). Combine dichloromethane layers and dry over anhydrous magnesium sulphate. Remove dichloromethane by distillation and purify the crude residue by column chromatograph on silica gel (ethyl acetate/hexane, 1:1) to afford 2-phenyl-*N*-(*p*-toluenesulfonyl)acetamidine as a white solid (m.p. 115-116 °C) ; ^{1}H NMR (400 MHz, CDCl$_3$) 8.02 (br, 1H), 7.75 (d, J) 8.2 Hz, 2H), 7.30-7.16 (m, 7H), 5.97 (br, 1H), 3.58 (s, 2H), 2.38 (s, 3H);   IR (KBr, cm^{-1}) 3327, 3328, 3238, 1654, 1545, 1449, 1414, 1277, 1103, 801.

Synthesis of aspirin

Aspirin is a medicine prepared by acetylation reaction of salicylic acid with acetic anhydride in the presence of sulphuric acid (equation 4.22).

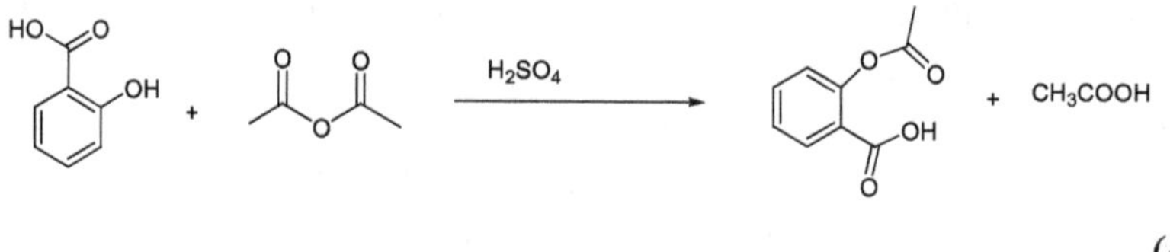

.....(4.22)

Procedure

Take salicylic acid (2 g, 0.014 mol) and acetic anhydride (2.8 ml, 0.027 mol) in a conical flask (25 ml). Add concentrated sulphuric acid (about 4 drops) to this solution and heat the reaction mixture on a water bath to about 50-60 °C. Allow the reaction mixture to cool and add water (15 ml) to this. The precipitate of aspirin will appear. Filter the precipitate and recrystallise the solid by dissolving it in a mixed solvent ethanol and water (3 ml: 7 ml).

Aromatic nitration

Nitration is a common electrophilic substitution reaction. In these reactions nitrating mixtures are conventionally prepared from sulphuric acid and nitric acid which generates nitro cation. The nitro cation is attacked by an electron rich aromatic ring, which leads to a substitution reaction. The position of nitration in the ring is decided by substituent on the ring. The *ortho* and *para* or *meta* directing effect becomes a dominating factor in these reactions. However, there are many nitrating agent, for example, some metal nitrate salts are capable of either

$$\text{.....(4.23)}$$

catalyzing or directly nitrating an aromatic ring. Let us consider nitration of benzaldehyde (equation 4.23). In this case aldehyde is a *meta* directing group, thus, a *meta* substituted product is formed. Unlike bromination introduction of nitrate groups deactivates aromatic ring, so second nitration in the ring becomes less pronounced.

Procedure

Take a mixture of sulphuric acid (25 ml, 32N) and fuming nitric acid (2.2 ml, 18N) and cool it in an ice bath. Add benzaldehyde (6 ml, 0.056 mol) dropwise to the reaction mixture. Do not allow the temperature of the reaction mixture to rise above 5 °C, while adding reagents. Heat the reaction mixture to about 40^0C and pour the reaction mixture to finely crushed ice (10 g). Filter the reaction mixture through sintered glass crucible to get the product as yellow solid.

Aromatic Sulphonation

Sulphonation is a common aromatic substitution reaction. This reaction is very useful to make water soluble organic compounds as sulphonate groups are highly polar and they form salts easily. Aniline can be sulphonated by this reaction (equation 4.24)

$$.....(4.24)$$

Procedure for sulphonation

Take aniline (20 ml, 0.21 mol) and add concentrated sulphuric acid (40 ml, 32N) dropwise with constant shaking. Keep the reaction mixture cool during the addition of sulphuric acid. Heat the reaction mixture at 180-190 °C for 5 hours. Check the completion of the reaction by with drawing 2 drops of the reaction mixture to a sodium hydroxide solution (5 ml, 2M), where it will dissolve without leaving the solution cloudy. Cool the reaction mixture; add ice cold water (400 ml) and allow to stand for 10 minutes. Filter the reaction mixture through sintered glass crucible to get the product.

Reduction reactions

Reduction of carbonyl compounds can be done in many ways. However each of them has their own characteristic properties. They can form alcohol or hydrocarbon. Hydrazine hydrate can reduce a carbonyl compound to a hydrocarbon without formation of alcohol. Consider the reduction of acetophenone to ethylbenzene (equation 4.25), it can easily done by hydrazine under basic medium.

$$.....(4.25)$$

Procedure

Place acetophenone (3.5 ml, 0.03 mol), ethylene glycol (3.0 ml, 0.055 mol) hydrazine hydrate (1 ml, 0.031 mol) and potassium hydroxide pellets (4.0 g, 0.071 mol) in a 100 ml two-necked round bottom flask fitted with a reflux condenser; insert a thermometer supported in a screw capped adaptor in the side neck so that the bulb dips into the reaction mixture. Warm the mixture on a water bath. After dissolving the potassium hydroxide, heat the mixture under reflux for 1 hr. Distil the product until the temperature rises to 175 °C. Separate the upper hydrocarbon layer from the distillate and extract the aqueous layer twice with diethylether. Remove the diethylether to obtain ethylbenzene.

Reduction by sodium borohydride

Sodium borohydride is a widely used reducing agent. It can be considered as combination of sodium hydride and borane. It is used for various purposes. For example, generation of metallic particles, reduction of functional groups, and hydroboration reaction. It reacts with Iodine to liberate BH_3 and the liberated BH_3 can be trapped as triphenyl phosphine adduct.

$$NaBH_4 + I_2 \quad \rightarrow \quad NaI + BH_3$$

$$.....(4.26)$$

$$BH_3 + PPh_3 \quad \rightarrow \quad BH_3PPh_3$$

$$.....(4.27)$$

Various functional groups such as nitro groups, carbonyl groups are reduced by sodium borohydride under mild condition. The reductions are dependent on reaction condition. For example, under ordinary reaction conditions, it does not reduce a nitro group. But borohydride along with a transition metal catalyst it can be done. Thus, advantage of reactivity difference can be taken to selectively reduce a functional group over another. Such an example of reduction of aldehyde group without affecting a nitro group is the reduction of 4-nitrobenzaldehyde with sodium borohydride (equation 4.28).

$$.....(4.28)$$

Procedure

Place a solution of *p*-nitrobenzaldehyde (1.5 g, 0.01 mol) in methanol (10 ml) in a three necked flask (100 ml). Stir the slurry and add a solution of sodium borohydride (0.14 g, 0.004 mol sodium borohydride in 0.02 ml of 2M sodium hydroxide diluted with 1.8 ml water) dropwise keeping the temperature at 18-25 °C. Treat a small portion of the reaction mixture with sulphuric acid. Remove most of the methanol by distillation on a steam bath, and dilute the residue with water (100 ml). Extract the mixture with ether (100 ml) by a separatory funnel. Wash the upper layer with water. Distil the ether layer to obtain transparent oil. Further distil the residual pale yellow oil under reduce pressure. Collect the *p*-nitrobenzyl alcohol at 183-185 °C/17 mm Hg .

Oxidation reactions

Oxidation reactions are very common in organic chemistry. The choice of reagents for specific oxidation is of great interest. Commonly used oxidizing reagents include metal salts at higher oxidation state. The metal at higher oxidation state can get converted to lower oxidation state. The simplest oxidation reactions are the oxidation of alcohols to carbonyl compounds. Such reactions can be done by using pyridinium chlorochromate. The oxidation of heptane-1-ol is given in equation 4.29.

$$.....(4.29)$$

Oxidation of 1-Hexadecanol

Take pyridinium chlorochromate (3.23g, 15 mmol) in anhydrous dichloromethane (20 ml). Add heptan-1-ol (2.42 g, 10 mmol) in dichloromethane (2 ml) in one portion to the solution. After 1.5 hrs add dry diethylether (20 ml). Decant the supernatant solution from the black gum. Wash the insoluble gum thoroughly with three portions of anhydrous ether (50 ml). Take the combined organic solutions and pass it through a short pad of florosil. On removal of ether by distillation gives the 1-hexadecanal.

Preparation of pyridinium chlorochromate (PCC)

The reagent pyridinium chlorochromate is prepared by the reaction of chromium trioxide with pyridinium chloride as shown in equation 4.30. Pyridinium chloride is generated by reacting pyridine with hydrochloric acid. The compound is hygroscopic and it dissolves in water very easily. Traces of moisture present makes it to decompose and it turns brown on leaving out side desicator in open condition.

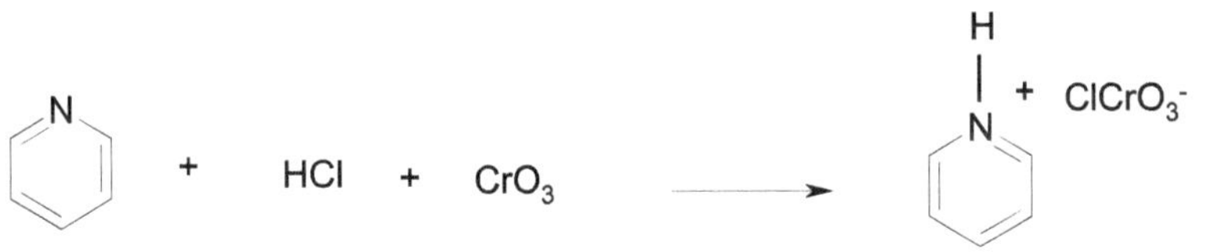

$$\ldots\ldots(4.30)$$

Procedure

Add hydrochloric acid (10.1 ml, 6M) to chromium(VI) oxide (10 g, 1 mol) with stirring in a conical flask. After 5 minutes cool the homogeneous solution to 0 °C and carefully add pyridine (7.91 g, 1 mol) over 10 minutes while cooling the solution to 0 °C. A yellow orange solid will be formed. Collect the solid on a sintered glass funnel and dry under vacuum.

Rearrangement cum oxidation by base

Benzil can be converted to benzilic acid by reacting it with potassium hydroxide (equation 4.31).

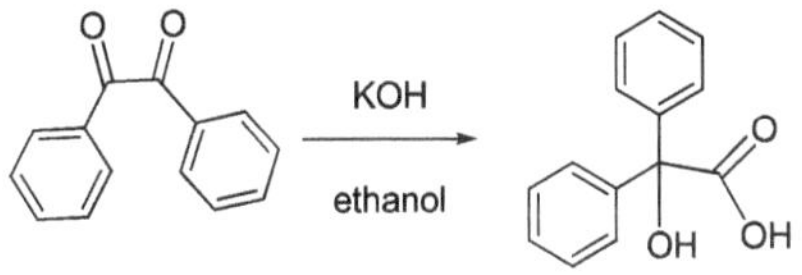

$$\ldots\ldots(4.31)$$

The reaction involves attack of a hydroxide at the carbonyl followed by migration of a phenyl group as illustrated in scheme 4.8.

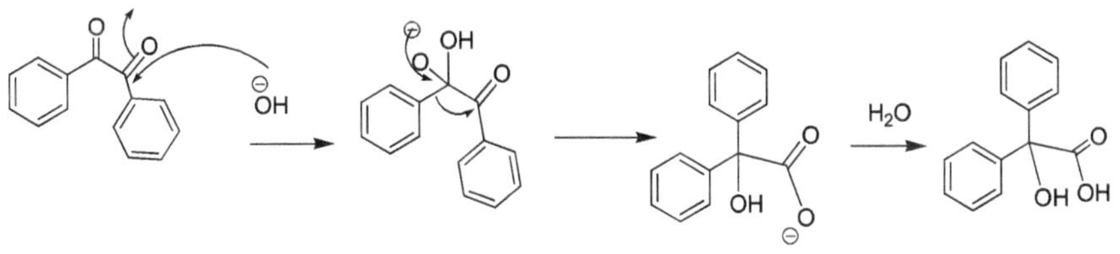

Scheme 4.8 The oxidative rearrangement of benzil

Procedure

Take benzil (4 g, 19 mmol), potassium hydroxide (5 g, 89 mmol) in water (10 ml) and ethanol (10 ml). Reflux the solution for 15 minutes.

Pour the reaction mixture to 100 ml ice cold water. To the solution add charcoal (1 g) and filter the charcoal. Take about 75 g of crushed ice and add to the filtrate followed by concentrated hydrochloric acid (15 ml). Benzilic acid will precipitate out, filter and dry (m.p. 149-151 °C)

Dihydroxylation reaction

Cis-dihydroxylation of alkenes by osmium tetroxide in an inert solvent gives selectively 1,2-diols. Cyclohexene is converted to cis-cyclohexane-1,2-diol by using osmium tetroxide and catalytic amount of hydrogen peroxide in t-butyl alcohol as shown below (equation 4.32).

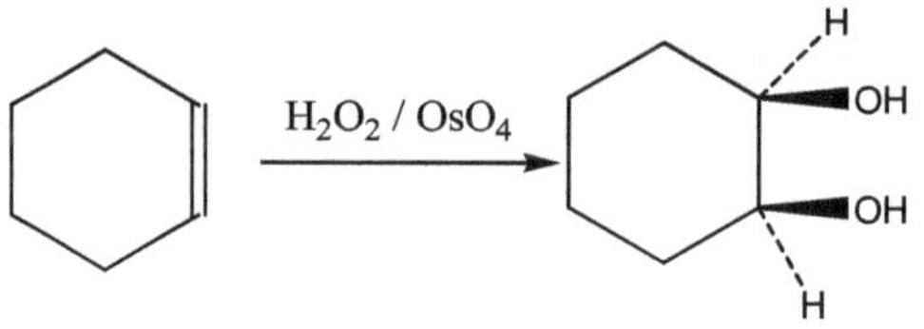

$$\text{H}_2\text{O}_2 \,/\, \text{OsO}_4$$

.....(4.32)

Procedure

Take a mixture of t-butyl alcohol (100 ml) and hydrogen peroxide (25 ml, 30 %), separate it to two layers by adding anhydrous sodium sulphate. Remove the alcohol layer and dry it over anhydrous magnesium sulphate. The resulting liquid is the catalyst solution which contains hydrogen peroxide in t-butylalcohol. Add cyclohexene (8.2 g, 0.1 mol) and osmium tetroxide (3 ml, 0.3 % solution in t-butyl alcohol) to the catalyst solution. Cool the solution to 0 °C and allow it to stand overnight. Remove the solvent and the unreacted any cyclohexene from the reaction mixture by distillation. Fractionate the residue and collect the fraction having boiling point 120-140 °C at 15 mm Hg pressure. The distillate solidifies on standing.

Oxidation by selenium dioxide

Selenium dioxide is a very selective oxidizing agent for an alkyl group adjacent to a carbonyl group. The methyl group adjacent to carbonyl gets oxidized to aldehyde group and a dicarbonyl compound is formed. Let us consider oxidation of acetophenone to phenylgyloxal by seleniumdioxide. The reaction is represented in equation 4.33.

$$\underset{\underset{Ph}{}{\overset{O}{\parallel}}}{C}-Me \xrightarrow{\text{SeO}_2} \underset{Ph}{\overset{O}{\parallel}}C-\underset{H}{\overset{O}{\parallel}}C + Se + H_2O$$

$$.....(4.33)$$

Procedure

Take selenium dioxide (55.5 g, 0.5 mol) in water (10 ml) and dioxan (300 ml) dissolve it by heating at 50-60 °C. Add acetophenone (60 g, 0.5 mol) to the solution and reflux the reaction mixture for 4 hrs with stirring. Decant the hot solution from the precipitated selenium. Distil off the dioxan and water from the reaction mixture. Collect the product phenylglyoxal by distillation under reduced pressure at 95-97 °C.

Oxidation of table sugar to oxalic acid

Chemically ordinary table sugar is a simple carbohydrate namely sucrose; it is also known as saccharose. Sucrose can easily be oxidized to oxalic acid. Such oxidation can be caused by concentrated nitric acid as follows:

Sucrose $\xrightarrow{\text{Conc. HNO}_3}$ Oxalic acid

$$.....(4.34)$$

Oxidation procedure: (The experiment should be carried out in a fume cup board). Take cane sugar (10g, 29 mmol) in a round bottomed flask. Add concentrated nitric acid (50 ml) to this and place it over a water bath to start reaction. After warming exothermic reaction will start and allow the reaction to boil of its own, heat. If it stops boiling heat it to boiling temperature and boil for two hours. Brown fumes will appear. Keep the reaction mixture at room temperature. Oxalic acid will appear as white crystals, collect the crystals. Find out the melting point (m.p. 102 °C)

Diazotisation reactions

Diazo compounds are used in preparing dye. They are also used in various organic transformations. They can lead to radicals and thus can form C-C bonded compounds easily. They serve as test for aromatic amines. They can be easily prepared by reactions of aromatic amine with sodium nitrite under acidic condition. Let us consider the preparation of a color indicator namely para-red via diazotization reaction. The reaction sequences are shown in equation 4.35 and 4.36.

$$\ldots\ldots(4.35)$$

$$\ldots\ldots(4.36)$$

Preparation of *p*-nitrobenzenediazonium chloride

Prepare a solution of concentrated sulphuric acid (2 ml) in water (10 ml). Place *p*-nitroaniline (1.4 g, 0.01 mol) in a beaker and add the above sulphuric acid solution to it, heat the mixture gently to make a homogeneous solution. Cool the mixture to about 10 °C by placing the reaction flask in an ice-bath. Make a solution of sodium nitrite (0.69 g, 0.01 mol) in water (2 ml) in another beaker and cool this also in an ice bath. Add the cold solutions of sodium nitrite drop wise to the ice cooled reaction mixture such that the temperature does not rise above 10 °C. An intense red solution will be formed. Store the resulting solution of *p*-nitrobenzene diazonium sulphate in an ice bath.

Prepare a solution of β-naphthol (1.44 g, 0.01 mol) in sodium hydroxide solution (25 ml of 10%) and store it at 10 °C. Add the diazonium salt solution to this solution dropwise with constant stirring. Stir the solution for 10 mins. Acidify the mixture with hydrochloric acid (1N), a red solid will be formed. Filter the solid; dry and recrystallise it from toluene.

Phenylhydrazone of carbonyl compounds

Aliphatic and aromatic compounds undergo nucleophilic addition reactions with different amines and hydrazine derivatives. These reactions are used for ligand synthesis and for characterization of carbonyl compounds. They are also one of the most important reactions in biology as the amino acids can combine with a carbonyl group via these types of addition reactions. The reaction of phenylhydrazine with carbonyl compounds leads to phenylhydrazones (equation 4.37).

Hydrazone derivative

$$\ldots\ldots(4.37)$$

Preparation of phenyl hydrazone

Take aldehyde or ketone (100 mg) in methanol (4 ml) with four drops of phenylhydrazine. Boil the mixture for 1 min and add 1 drop of glacial acetic acid, and boil the mixture gently for 3 mins. Add cold water dropwise and cool the solution which will lead to crystallization of phenylhydazone if they are solid. Collect the crystals and recrystallise from methanol.

Oximes

Oximes are important class of compounds derived from condensation of carbonyl compounds with hydroxyl amine. They are widely used as ligand with different metal ions and are used as protecting group of alcohols and also for preparation of other functional groups such as amide and cyanide. They can be prepared by reaction of a carbonyl compound with hydroxylamine in the presence of base. In many reactions hydroxylamine hydrochloride is used as the source of hydroxylamine (equation 4.38). The oximes undergo rearrangement reaction in presence of strong acids or Lewis acids to give amides and the reaction not only has synthetic value but is important from mechanistic point of view.

$$\ldots\ldots(4.38)$$

Benzophenone oxime can be prepared from the reaction between benzophenone and hydroxylamine in the presence of sodium hydroxide. Benzophenone oxime on reaction with phosphorus pentachloride gives benzanilide (equation 4.39). It is a commercially important reaction, since the caprolactum is produced via rearrangement reactions of oxime. Caprolactum is the main feedstock for the production of nylon-6.

$$\text{Ph}_2\text{C=N-OH} \xrightarrow{\text{PCl}_5} \text{PhCO NH Ph}$$

.....(4.39)

Procedure for preparation of benzanilide

Dissolve benzophenone oxime (2 g, 17.6 mmol) in anhydrous ether (20 ml) and place it on an ice bath; to this add powdered phosphorus pentachloride (3 g, 14.4 mmol). Distil the solvents and add water (25 ml). Warm the reaction mixture to obtain a white solid. Filter the white solid and recrystallise the product from ethanol.

Protection of carbonyl groups

When multiple functional groups are present or to carry out a specific reaction over another reaction some functional groups need to be converted to less reactive group. However, attention is to be given such that the conversion carried out should be such that after the desired transformation, by another reaction the transformed group can be recovered. Such process is called protection of a functional group. For example alcohol of a hydroxyl and aldehyde containing compound can be selectively oxidized by protecting the aldehyde group. If precaution is not taken aldehyde group will get oxidized. At the end of oxidation the aldehyde group can be recovered. It is possible to protect a reactive carbonyl group selectively in the presence of a less reactive functional group. The most useful protections are done through formation of corresponding the acyclic and cyclic acetals or ketals, and the acyclic or cyclic thioacetals or ketals. The protective group may be removed by reacting it with an aqueous acid, giving back the original carbonyl. This step is called deprotection. The protection of carbonyl group in ethyl acetoacetate with ethane-1,2-diol in the presence of *p*-toluene sulphonic acid is shown in Scheme 4.9.

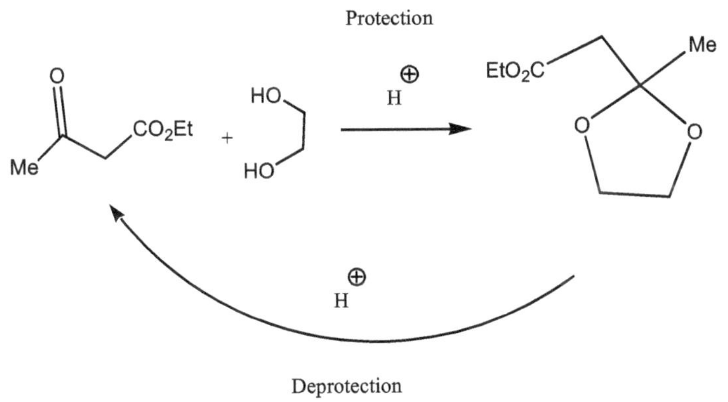

Protection

Deprotection

Scheme 4.9 An example of a protection and deprotection reactions

Procedure

Add ethyl acetoacetate (30 g, 0.23 mol), ethane-1,2-diol (16 g, 0.26 mol), *p*-toluene sulphonic acid monohydrate (0.1 g, 5 mmol) to benzene (50 ml) in a flask fitted with a Dean and Stark water separator and a reflux condenser. Reflux the reaction mixture for 6-8 hrs so that benzene with water is collected in the side arm. On removal of the solvent the solid product of cyclic acetal will be obtained. Recrystallise the acetal from dry methanol. Deprotection of the acetal can be done by refluxing an aqueous methanolic solution of acetal with sulphuric acid (1N).

Grignard reagent

Grignard reagent is one of the most widely used reagents in organic synthesis. Almost all types of organic compounds can be synthesized by using Grignard reagents. It is especially useful for formation of new carbon-carbon bonds. The Grignard reagent is prepared from the direct reaction of magnesium with an alkyl halide. Many a time iodine is used as activating agent for preparation of Grignard reagent. Let us take an example of use phenyl magnesium bromide to prepare benzoic acid. Phenyl magnesium bromide reacts with carbondioxide and benzoic acid can be prepared from this reaction. (Scheme 4.10)

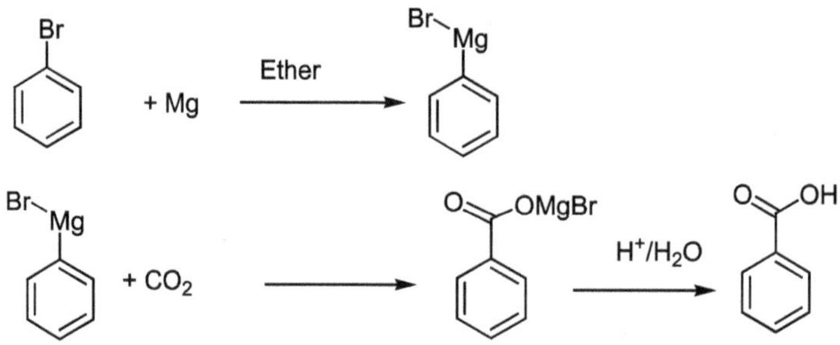

Scheme 4.10 Synthesis of benzoic acid from bromobenzene

Preparation of phenyl magnesium bromide

Take finely cut magnesium turnings (2.4 g, 0.1 mol) in dry diethylether (15 ml) in a two necked round bottomed flask fitted with reflux condenser and dropping funnel. Place the entire set up under continuous flow of nitrogen gas. Add bromobenzene (10.5 ml, 0.1 mol) from the dropping funnel dropwise to the reaction mixture. Warm the mixture to get turbidity in the mixture. Add additional 15 ml of dry ether to the reaction mixture and add the remaining bromobenzene dropwise over a period of 15 minutes. After completion of the addition, reflux the mixture for a period of another 10 minutes. Cool the reaction mixture to room temperature. The solution contains phenyl magnesium bromide. During formation of Grignard reagent, dissolution of magnesium takes place, and it can be visibly seen. Add finely crushed solid carbon dioxide (dry ice 25-35 g) to the reaction mixture. While solution is cooled, add slowly 75 ml of ether with gentle stirring. Allow the excess carbon dioxide to get evaporated by bringing the solution to room temperature. Add hydrochloric acid (10 ml, 1N) and extract the ether layer. On evaporation of ether layer gives benzoic acid as white solid.

Carbenes

Carbenes are electron deficient 6-electron highly reactive species. It is generally derived from trihalogen compound by treating with strong base. However, decomposition of a diazonium compounds under photolytic condition leads to carbenes. Depending on the spin multiplicity they are two types, singlet and triplet. The singlet carbene has pair of electron in one sp2 hybrid orbital with a vacant p-orbital whereas a triplet carbene has two unpaired electron in two p-types of orbitals (figure 4.14). The singlet carbenes easily reacts with oxygen which also has a pair of unpaired electrons. Thus, from a mixture of singlet and triplet carbene, singlet carbene can be removed easily.

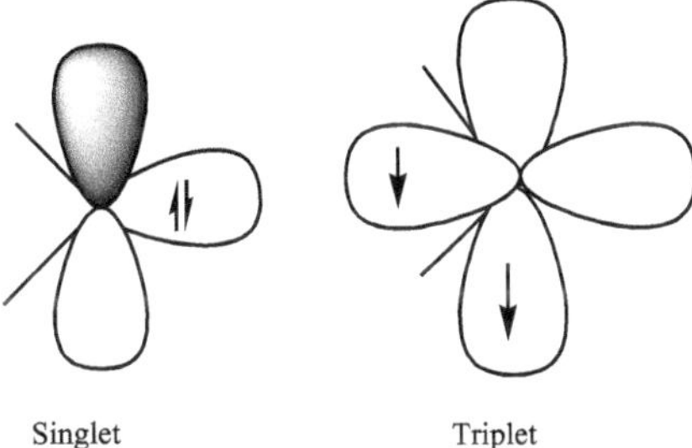

Fig. 4.14 Structure of singlet and triplet carbenes

The singlet carbenes add to an olefin in a stereo-specific manner, while triplet carbene addition to addition to olefin is non specific. This happens as singlet carbene can directly bind to two paired electron of the π-bond whereas the triplet binds to them through radical manner.

$$\ldots\ldots(4.40)$$

$$\ldots\ldots(4.41)$$

Reimer-Tiemann reaction

Phenol is converted to salicylaldehyde by using a base and chloroform in Reimer-Tiemann reaction. Carbene intermediates are involved in such reactions.

$$\ldots\ldots(4.42)$$

$$OH^- \quad H\text{-}CCl_3 \longrightarrow H_2O + \overset{\ominus}{:}CCl_3^- \longrightarrow Cl^- + :CCl_2$$

Scheme 4.11 Carbene generation from chloroform

Dichlorocarbene, a carbene formed in the reaction between chloroform and strong base. Dichlorocarbene reacts in the *ortho-* and *para*-positions, but more preferably at *ortho*-position of sodium phenoxide to give the dichloromethyl substituted phenol and subsequent basic hydrolysis of the dichloro-derivative gives the desired aldehyde.

Preparation of anthralinic acid

An amide on reaction with sodium hypobromite or hypochlorite gets converted to a primary amine. The hypobromite can be generated in situ by reaction of bromine with sodium hydroxide. The anthralinic acid can be prepared from phthalimide by such reactions. The reaction involves three steps. In the first step phthalimide ring is opened by reaction with sodium hydroxide and in the next step the amide groups get converted to amine group and sodium salt of anthralinic acid is formed. In the last step the salt formed is protonated to give anthralinic acid. The reaction is shown in scheme 4.12

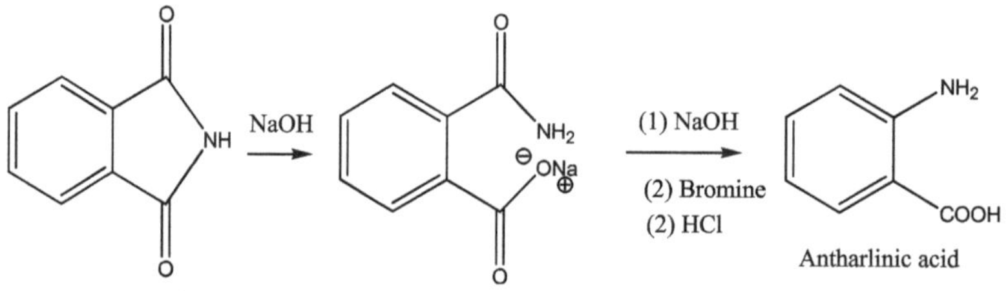

Scheme 4.12 Degradation of an amide to an amine

Procedure

Take an ice cooled solution of sodium hydroxide (30 g, 0.75 mol) in water (120 ml). Add bromine (26.2 g, 0.33 mol) in one portion to this solution. Add finely powdered phthalimide (24 g, 0.16 mol) to the cold reaction mixture with stirring. Add a solution of a sodium hydroxide (22 g in water 80 ml) to the reaction mixture. Heat the solution to about 80 °C for 2 min. Cool the solution and add concentrated hydrochloric acid until the solution becomes neutral. Precipitate the anthranilic acid by addition of glacial acetic to the solution and filter the solid and wash with cold water. Recrystallise the product from hot water.

Sigmatropic rearrangement

The 3,3 sigmatropic rearrangement reactions are common in conjugated olefinic compounds. Such reactions occur in systems having structural features of allyl vinyl ether. For example, allyl phenyl ether undergoes rearrangement reaction to give 2-allyl phenol and 4-allyl phenol. To make such reactions specific the 4-position of the aromatic ring is to be blocked. For example 3-allyl-4-hydroxy benzaldehyde undergoes reaaarngemet to give 2-allyl 1-hydroxy benzaldehyde (equation 4.43).

$$\text{.....(4.43)}$$

Procedure

Take 4-allyloxy benzaldehyde (7 g, 0.052 mol) in a round bottom flask (50 ml capacity). Add N, N-dimethylaniline (5 ml) to it. Reflux the solution on an oil bath at 200 °C for six hours. Treat the reaction mixture with few drops of hydrochloric acid, followed by addition of water (10 ml) and ethylacetate (30 ml). By such treatment N, N-dimethyl aniline goes into the aqueous layer by forming the corresponding ammonium salt. Extract the product with ethylacetate. Collect the ethyl acetate layer and dry it over anhydrous sodium sulphate. Concentrate the ethylacetate solution. Purify the crude product by column chromatography.

Photochemical reactions

Photochemical reactions in organic chemistry find an important place. The 2 + 2 cycloaddition of olefinic compounds are well studied reactions to make four member rings, namely cyclobutane rings. The photochemistry of carbonyl compounds are interesting as depending on the presence of α-hydrogen or not the reactions varies. Moreover, several aromatic ketones are used as sensitizer in photochemical reactions. Isopropanol can be photooxidised to ketone. The oxidation is effectively done by soluble copper(II) complexes; which in turn gets reduced to metallic powder. Benzophenone in this reaction acts as a sensitizer (equation 4.44).

$$\text{(iso-propanol)} \xrightarrow[\substack{\text{benzophenone} \\ \text{light}}]{\text{Cu(AcAc)}_2} \text{(acetone)}$$

.....(4.44)

Photochemical oxidation of iso-propanol by copper (II) acetylacetonate

Take a solution of copper (II) acetylacetonate (0.05 g, 0.19 mmol) and benzophenone (0.112 g, 0.64 mmol) in isopropyl alcohol (100 ml) in a quartz tube and photolyse the solution for 1 hr by using a lamp (630 nm). After I hr copper suspension will be formed and filter off the copper formed. Add 2,4-dinitro phenylhydrazine solution to the filtrate. A yellow precipitate will be obtained. Filter and recrystallise the solid from ethanol/water. Characterize it as 2,4-dinitro phenyl hydrazone derivative of acetone (m.p. 124-126.5 $^\circ$C).

Nylon polymer

Nylon is an amide based polymer commercially used for fabric material. It is prepared by simple substitution reaction on acid dichloride of a dibasic acid with diamine in presence of a base (equation 4.45).

$$Cl-C(=O)-[CH_2]_8-C(=O)-Cl \;+\; H_2N-[CH_2]_6-NH_2 \xrightarrow{\text{NaOH}} \left[-C(=O)-[CH_2]_8-C(=O)-NH-[CH_2]_6-NH- \right]_n$$

nylon

.....(4.45)

Procedure

Place a solution of hexamethylenediamine (0.5M, 50 ml in 0.5M sodium hydroxide) in a beaker (250 ml). Add slowly a solution of sebacoyl chloride (2 ml dissolved in 50 ml hexane) solution as a second layer on top of the diamine solution. While adding the acid chloride care should be taken to minimize agitation at the interface. A polymer film will be formed at the interface. With the help of a forcep, grasp the polymer film and pull it up and wind the polymer thread on a glass rod. Wash the polymer thoroughly with water or ethanol.

Phenol formaldehyde resin

Bakelite or phenol formaldehyde resin is one of the most important polymeric resin, which is used in our day to day life. It can be prepared by the reaction of phenol with formaldehyde in the presence of an acid or base (scheme 4.13).

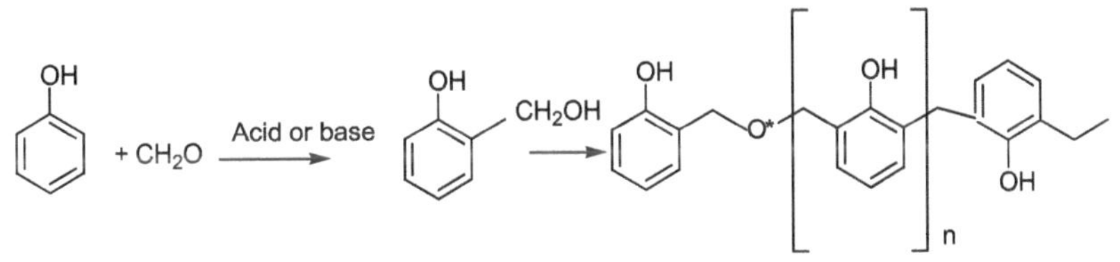

Scheme 4.13 Preparation of phenol-formaldehyde resin

Preparation of phenol-formaldehyde resin

Mix phenol (24 g, 0.5 mol), formalin (3 ml, 1 mol) and ammonia (0.25 ml) together in a round bottomed flask by placing it over an ice bath. Reflux the reaction mixture on an oil bath. An exothermic reaction takes place as the polymer forms. Care should be taken to control the temperature. Distil the water layer under vacuum by heating the mixture at about 75 °C. Continue distillation until the resin is clear. Pour the resin to a tray where it hardens to form solid.

A multistep synthesis

Synthesis of but-2-ynoic acid can be achieved in three steps from ethtylacetoacetate. First step involves formation of a heterocyclic compound. In the second step the heterocycle is brominated. Finally dehydrorbromination of the brominated compound leads to the desired product as illustrated in scheme 4.14.

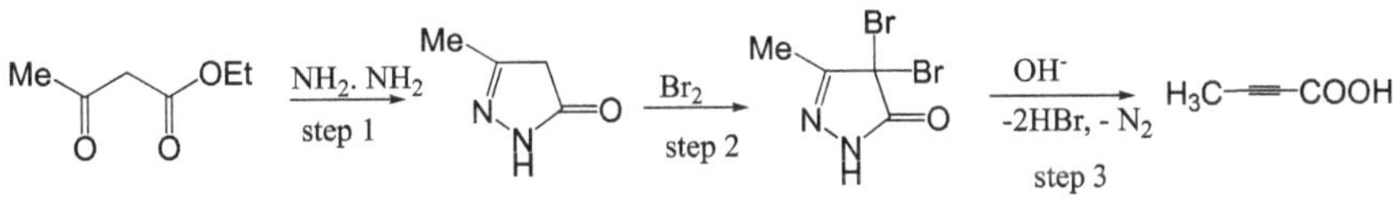

Scheme 4.14

Step 1: Preparation of 3-methylpyrazol-5-one

Place ethyl acetoacetate (6.5 g, 0.05 mol) in a conical flask and add hydrazine hydrate (2.5 g, 0.05 mol) taken in absolute ethanol (5 ml) dropwise with constant stirring. Maintain the reaction temperature at 60 °C with further stirring for 1 hr. Cool the reaction mixture in an ice bath to complete the crystallization. Filter the crystals of 3-methylpyrazol-5-one and wash it with ice-cold ethanol.

Step 2: 4,4-Dibromo-3-methylpyrazole -5-one

Dissolve 3-methylpyrazol-5-one (2 g, 0.02 mol) in glacial acetic acid (10 ml) and add bromine (3.2 g, 0.02 mol) in glacial acetic acid (5 ml) and stir for five minutes. Add another solution of bromine (3.2 g in acetic acid 5 ml) to the above solution and leave the reaction mixture at ambient condition for overnight. Add water to precipitate 4,4-dibromo-3-methylpyrazole-5-one and filter. Wash the product with water (5 ml) and dry it in open air (m.p. 130-132 °C).

Step 3: But-2-ynoic acid

Take a solution of sodium hydroxide (2 g in 50 ml water) in a round bottomed flask and place the flask in ice bath to maintain the temperature at 5 °C. Add 4,4-dibromo-3-methylpyrazole-5-one (3.4 g, 0.01 mol) in portions wise to the solution and stir. An orange red solution will be formed. Stir the reaction mixture for 1 hr. Cool the reaction mixture and acidify with concentrated hydrochloric acid till the solution becomes acidic. Extract the acidified solution with ether about 500 ml. Upon distillation orange oil will be obtained. Take the orange oil and add petroleum ether to make homogeneous solution. Upon concentration to about 5 ml white crystalline product will appear (m.p. 74-75 °C).

Synthesis of a deuterated compound

Cis-3-D-allyl alcohol can be prepared by the reaction of propargyl alcohol with lithium aluminum hydride in diethyl ether followed by addition of deuterium oxide. This reaction must be carried out in fume cupboard and under complete anhydrous condition.

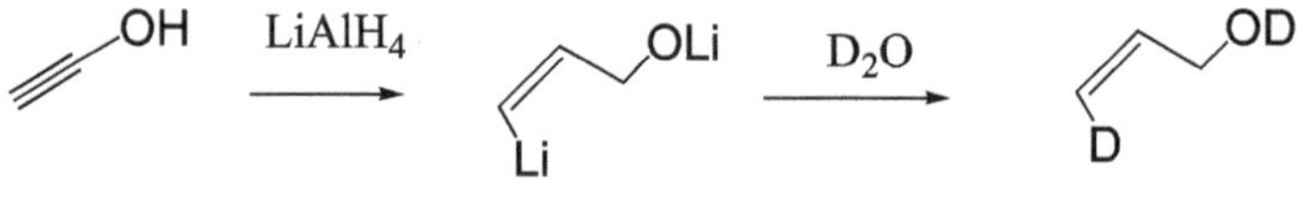

Scheme 4.15

Procedure

Take proporgyl alcohol (0.56 g, 10 mmol) in dry ether (200 ml) and place the reaction mixture in a round bottomed flask connected to a double wall condenser over ice bath. Add lithium aluminum hydride (0.41 g, 11 mmol) in portions so that the temperature does not rise above 10 °C. Allow the reaction to continue for two hours. Add drop wise deuterium oxide (0.5 ml) very carefully such that the temperature does not rise. To the reaction mixture add 2 ml dry ethylacetate slowly dropwise over a period of half an hour under ice cold condition. To this solution add ethanol (5 ml) and bring it to room temperature. Add 5 ml water and extract the organic layer. On removal of solvent by distillation will give the *cis*-deuterated alcohol.

Synthesis of heterocyclic compound

5-methylisoxazol-3-ol

It is a simple five member heterocyclic compound, prepared by simple condensation reaction of hydroxylamine with methylacetoacetate. This reaction take places in basic medium and initially the methylacetoacetate gets converted to amide derivative of hydroxylamine, which cyclises to give the product.

5-methylisoxazol-3-ol

$$.....(4.46)$$

Procedure

Take a solution of sodium hydroxide (2M, 50 ml) in water ice cold condition. Add methylacetoacetate (6.9 g, 0.5 mol) and hydroxylamine hydrochloride (3.85 g, 0.55 mol). Stir the mixture vigorously for 30 mins and pour the resultant solution in an ice cold solution of hydrochloric acid (30 ml, 2M). Leave the solution in refrigerator overnight and next day filter the solution to obtain the desired product (m.p. 101 °C).

CHAPTER **5**

Contemporary Experiments of Material Chemistry

Applied Materials

Introduction

The central point in chemistry to design useful and applied materials through easy synthetic procedure will continue to evolve. In this regard the construction of macromolecular system through supramolecular chemistry has been a central point of interest. A multi-component system of atoms, ions and/or molecules held together by non-covalent interactions such as hydrogen bond, Van der Waals forces, pi-pi interactions, and/or electrostatic effects makes a supramolecular assembly. These various interactions used to make supramolecular assemblies are far weaker than conventional chemical bonds; therefore supramolecular assemblies are less stable than molecular compounds. Whether it is the biology or chemistry, study of bio-inspired molecules would continue to attract interest. The material design will need of synthetic skill with imagination and the supramolecular chemistry along with nano-technology experiments is useful in modern day chemistry. Whether it is a light emitting diode, molecular wires, switches, rectifiers, storage materials all can be developed with the sound knowledge of synthesis coupled with the skill of physical measurement. Thus, the interdisciplinary work provides a better solution for preparation of useful materials. The following topics deals with some of the issues that have relevance to emerging material design.

Inclusion Compounds

The host–guest or supramolecular compounds started way back 1823 when Faraday reported the preparation of the chlorine clathrate hydrate. Subsequently graphite intercalates (1841), hydroquinone-hydrogensuphide clathrate (1849), the cyclodextrin inclusion compounds (1891), the preparation of the nickel cyanide ammonia inclusion compound with benzene (1897) carried a big message to the next century and in last 100 years study on these have led to develope independent subjects.

Clathrates are single phase solid derived from two components, one is called host and other is guest. The guest molecules are the one that are retained in the cavities created by host molecules. Hydroquinone has extensive hydrogen bonded structure and form different types of clathrate compounds. Hydroquinone forms two types of clathrates, they are known as α and β-form. Three hydroquinone molecules assemble together to act as host for smaller molecules in the α- form; whereas in β-form has six hydroquinone molecules assembles to encapsulate a guest molecule. Sulphur dioxide can also form clathrate with the hydrogen-bonded network of hydroquinone. Passing sulphur dioxide to a hydroquinone solution as per the following equation 5.1 forms a 1:1 molecular complex.

$$\text{.....(5.1)}$$

Hydroquinone clathrate with sulphurdioxide

Prepare a saturated solution of hydroquinone (5 g in 50 ml water) heated to 50 °C and to this solution pass a slow stream of sulphur dioxide gas for 30 minutes. Yellow precipitate will be obtained. Filter the precipitate through Buchner funnel and dry it by pressing between filter papers.

Molecular recognition

The N,N'-*bis*(glycinyl)pyromellitic diimide encapsulates fused aromatic hydrocarbons such as anthracene, phenanthrene and perylene. The carboxylic units of N,N'-*bis*(glycinyl)pyromellitic diimide are responsible for the formation of supramolecular assemblies through hydrogen bonding. Preferential inclusion of anthracene over phenanthrene in the supramolecular assembly of N,N'-*bis*(glycinyl)pyromellitic diimide occurs leading to isomeric separation. This process is based on pure aromatic π–π stacking interaction. The scheme 5.1 represent a series of inclusion compounds of N,N'-*bis*(glycinyl)pyromellitic diimide and Fig. 5.1 represents one crystal structure of an inclusion complex

An example of guest inclusion

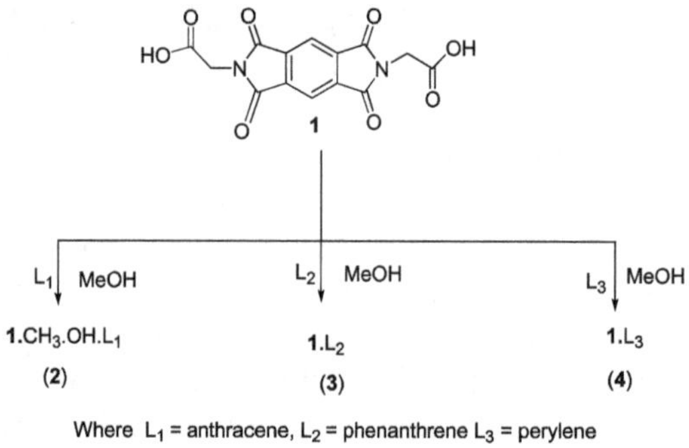

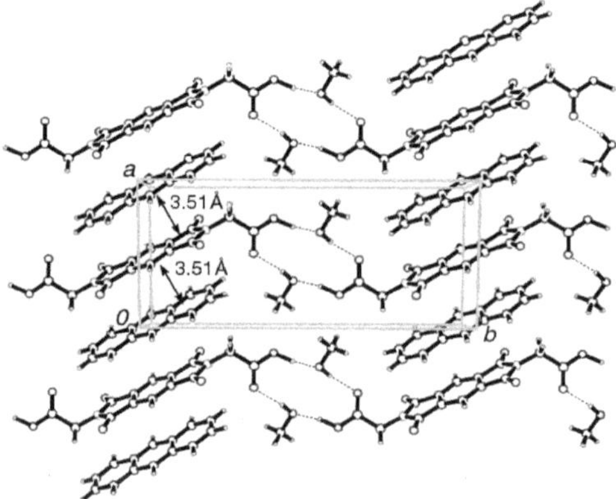

Scheme 5.1 1 is N,N′-*bis*(glycinyl)pyromellitic diimide

Fig 5.1 The crystal structure of N,N′-*bis*(glycinyl) pyromellitic diimide with anthracene

The parent carboxylic acid has a network structure through conventional hydrogen bonding between two units whereas in the inclusion complex with anthracene it has hydrogen bonding interactions

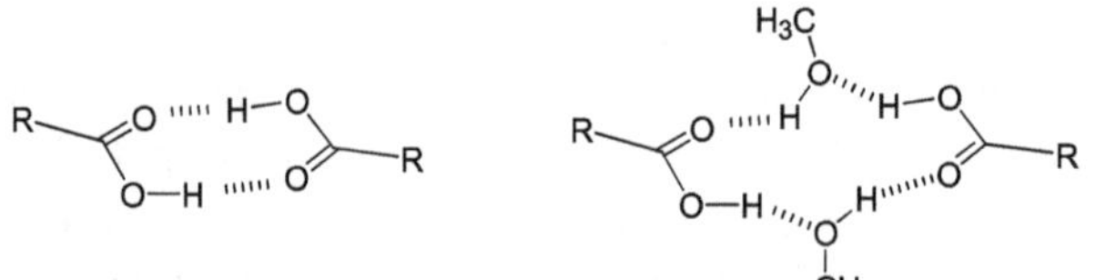

Hydrogen bonding pattern in
perylene and phenathrene
adduct of N,N'-bis(glycinyl)
pyromellitic diimide

Hydrogen bonding pattern in
Anthracene adduct of N,N'-bis(glycinyl)
pyromellitic diimide and methanol

Fig. 5.2

through two methanol molecules (Fig. 5.2). However, the conventional hydrogen bonding pattern among the carboxylic acid group is observed in the case of inclusion compound of N,N'-*bis*(glycinyl)pyromellitic diimide with phenanthrene and perylene. The difference between the solid state assemblies clearly suggests that there is role of solvent in such molecular recognition processes. The compound N,N'-*bis*(glycinyl)pyromellitic diimide is easily prepared by condensation reaction of pyromellitic dianhydride with glycine in acetic acid.

Synthesis of N,N'-*bis*(glycinyl)pyromellitic diimide

Take pyromellitic dianhydride (1.09 g, 5 mmol) in glacial acetic acid (30 ml) and make a solution by warming to 40 °C. Add finely powdered glycine (0.75 g, 10 mmol) to the reaction mixture and reflux the solution for 4 hrs. A yellow solution will be obtained. Concentrate the solution under reduced pressure and to obtain colorless solid. IR (KBr, cm^{-1}): 3344 (bs), 1787 (s), 1736 (s), 1418 (s), 1321 (s), 1127 (s). ^{1}HNMR (DMSO-d^{6}) 8.3 (s, 2H), 4.3 (s, 4H), 3.4-4.2 (b, 2H).

Anion recognitions

Halide ions play a major role in environment and biology. Chloride ion channels are found in biological molecules. The crystal structures of two chloride ion channels obtained from S.typhimurium and E.coli contain identical ion pores, and in both cases the chloride ions are held by four hydrogen bonds as illustrated in Fig. 5.3. Anion recognition is an active area of research in supramolecular chemistry. One of the basic principle in anion binding is that there must be a site on a heteroatom either in the protonated or neutral form to form hydrogen bond with the anion and binding constant is good enough to hold an anion with a host molecule. One such anion binding is shown in Fig. 5.4.

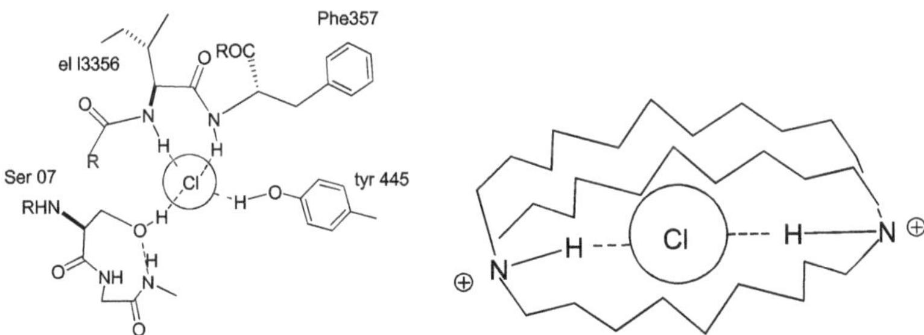

Fig. 5.3 Anion binding in biology   **Fig. 5.4** Anion binding to a receptor

Azacrowns are useful compounds for anion recognition as well as for detection of metal ion selectively. Synthesis of an aza-crown namely *N*-(8-Methoxy-2-methylquinolinyl) aza-12-crown-4 for such study is shown in scheme 5.2

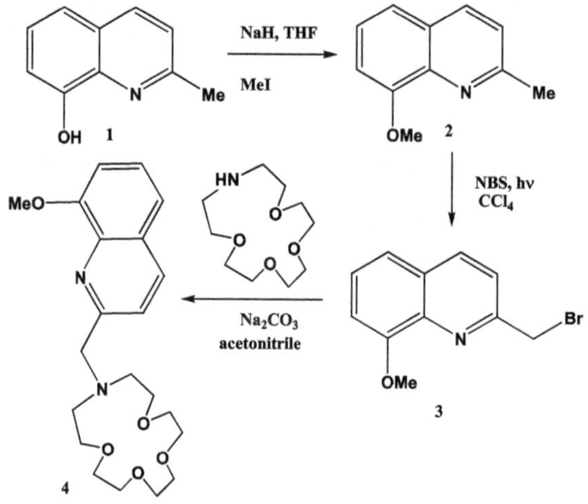

Scheme 5.2

Synthetic procedure

8-Methoxy-2-methylquinoline (2). Take sodium hydride (2.17 g, 54 mmol, 60% in mineral oil) in THF (150 ml). Add a solution of 8-hydroxy-2-methylquinoline (**1**) (7.96 g, 50 mmol) in 150 ml of THF. Stir the reaction mixture at room temperature for 2 h. To the reaction mixture, add methyliodide (7.67 g, 54 mmol). Reflux for 72 hrs and evaporate the solvent under reduced pressure to obtain a white residue. Dissolve the residue in dichloromethane (100 ml). Wash the organic part with 5% aq sodium hydroxide (3 × 40 ml) and then with water (3 × 40 ml). Dry the dichloromethane over sodium sulphate and remove the solvent. Purify the product by column chromatography to obtain **2** as a white solid with mp 125 °C.

N-(8-Methoxy-2-methylquinolinyl) aza-12-crown-4 (4)

Make a solution of **2** (1.00 g, 5.8 mmol) and *N*-bromo succinimide (1.13 g, 6.4 mmol) in carbon tetrachloride (200 ml). While refluxing the solution irradiate it with a 500 watt tungsten lamp for 2 hrs. Filter the reaction mixture and evaporate solvent from the filtrate to give a solid. Purify **3** by column chromatography (m.p 116-118 °C).

Take a mixture of **3** (1.61 g, 6.8 mmol), aza-12-crown-4 (0.80 g, 4.6 mmoles), and sodium carbonate (1.45 g, 13.7 mmol) in acetonitrile (100 ml). Reflux the solution for 4 days. Filter the reaction mixture and evaporate the solvent. Purify the residue by column chromatography on alumina with ethylacetate. The compound **4** will be obtained as an oil. IR (neat, cm^{-1}): 1259, 1124, 1107. ^{1}HNMR (CDCl$_3$): 2.85 (triplet, 4H), 3.40-3.90 (multiplet, 12H), 3.95-4.25 (multiplet, 5H), 6.85-7.15 (multiplet, 1H), 7.20-7.50 (multiplet, 2H), 8.05 (doublet, 2H).

Trapping of unstable species

From biological and environmental point of view inclusion of water soluble anions such as phosphates, and sulphates are of interest. It has been shown by vibrational spectroscopy that methyl hydrogen sulfate can be generated in solution. The reaction of methanol with sulphuric acid can lead to methyl hydrogen sulphate and same reaction in the presence of pyridine, could either lead to pyridinium hydrogen sulphate salt as well as methyl sulfate salt as illustrated in equation 5.2 and 5.3. Methylsuphate anion can be trapped by salt formation reactions of sulphuric acid in methanol with pyridine bis-phenols namely 2-[*bis*(4-hydroxy-3,5-dimethylphenyl) methyl]pyridine and 4-[*bis*(4-hydroxy-3,5-dimethyl phenyl) methyl] pyridine (Fig 5.5). These are very simple systems to trap MeOSO$_3^-$ over the trapping of the MeOSO$_3^-$ anion by calix[4]arene derivatives or metal-complexes.

$$CH_3OH \;+\; H_2SO_4 \;\xrightarrow{\text{water}}\; CH_3SO_4H$$

Methyl hydrogen sulphate

$$.....(5.2)$$

$$CH_3OH \;+\; \text{(pyridine)} \;+\; H_2SO_4 \;\xrightarrow{H_2O}\; \text{(pyridinium)}\; HSO_4^- \quad or \quad \text{(pyridinium)}\; CH_3SO_4^-$$

$$.....(5.3)$$

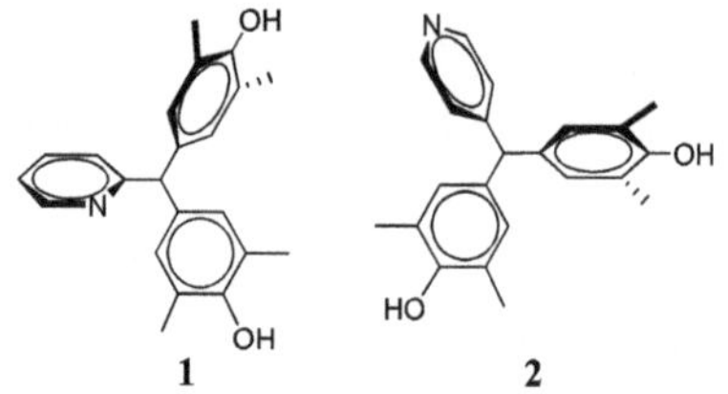

Fig. 5.5 Structure of 2-[*bis*(4-hydroxy 3,5-dimethylphenyl)methyl]pyridine (**1**) and 4-[*bis*(4-hydroxy 3,5-dimethylphenyl)methyl]pyridine (**2**)

Synthesis of 2-[*bis*(4-hydroxy-3,5-dimethylphenyl)methyl] pyridine

Take 2,6-dimethylphenol (0.49 g, 4.1 mmol) and pyridine-4-carboxaldehyde (0.213 g, 2 mmol) in dichloromethane (10 ml) and keep the mixture in an ice bath at 0 °C for 10 minutes. To this solution add trifluoro acetic acid (2 ml) over a period of 10 minutes. Keep the reaction mixture at room temperature for 24 hrs and add sodium bicarbonate to neutralize solution. Extract the organic part by dichloromethane (2 × 25 ml) and water (20 ml). Take the dichloromethane layer and dry it over sodium sulphate (anhydrous) and remove the solvent. Recrystallise the product from chloroform. IR (KBr, cm^{-1}): 3385 (s), 1634 (s), 1485 (s), 1147 (s). ^{1}HNMR (DMSO-d^6): 7.60 (2H, doublet); 6.70 (2H, doublet), 6.60 (4H, singlet), 5.43 (1H, singlet), 3.70 (2H, singlet), 2.18 (12H, singlet).

Calix-4-arene

A calixarene is a macrocycle or cyclic oligomer prepared from condensation of phenols with aldehydes. Calixarenes contains a hydrophobic aromatic ring with or without substituent and hydrophilic hydroxyl groups. The calix[4]arene (Fig. 5.6) has four phenolic rings. Some of the calix arenes are difficult to prepare, however, pyrogallol[4]arene can be easily prepared by mixing pyrogallol with aldehyde like isovaleraldehyde in the presence of a catalytic amount of p-toluene sulphonic acid in a mortar and pestle. Calixarenes finds applications in enzyme mimic, ion sensitive electrodes, sensors, non-linear optics. The calix arenes can have different conformers that can be stabilised by various cations. The reaction of pyrrogallol with isovaleraldehyde procceds at room tempertaure in the presence of an acid catalyst. The reactio leads to cyclic a product as illustrated in Fig. 5.7. This compound in absence of solvent self assembles to form nano-capsules.

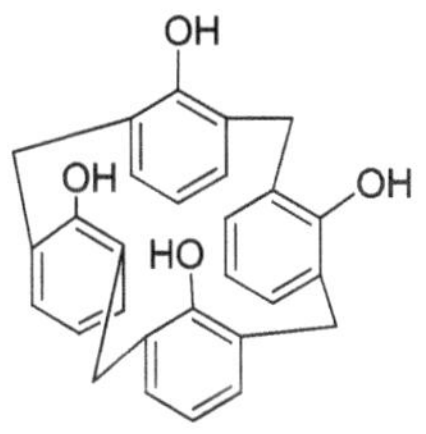

Fig. 5.6 Calix [4] arene

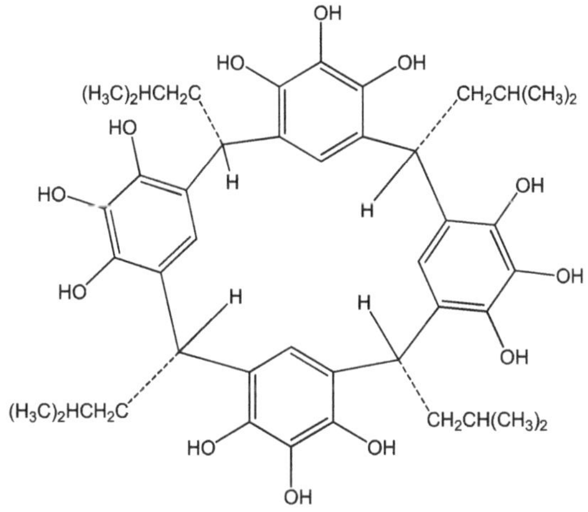

Fig. 5.7

Synthesis of C-isobutylcalix[4]pyrogallolarene

Add isovaleraldehyde (0.89 ml, 7.9 mmol) dropwise to a fine powder of pyrogallol (1 g, 7.9 mmol) in a pestel. To this mixture a catalytic amount of solid *p*-toluenesulfonic acid (50 mg, 0.3 mmol) with constant milling using a mortar and pestle. Brittle white solid will be formed within two minutes of grinding.

Construction of lamellar structure from *bis*-phenols

From *bis*-phenols lamellar structures can be constructed by use of weak interactions. The hydroxy group or the amino groups can serve as donor as well as acceptor to hydrogen bonds and it is possible to obtain extended networks. They may be represented by triangular representation as illustrated in Fig. 5.8. Combining the donor-acceptor sites a lamellar structure can be constructed. However, changing the amino group with a nitro or aldehydic groups with similar concept layered structures can be generated.

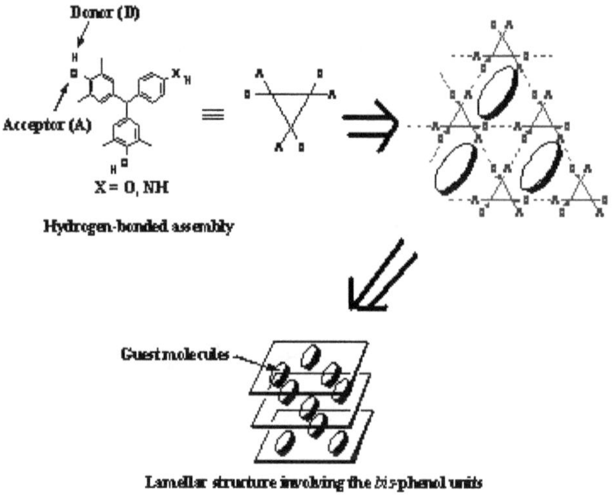

Fig. 5.8 Lamellar structures generated fro bis-phenols

Catalytic systems

Amino acids have been useful from biology and chemistry. It has chiral centers that can directly be used for chiral ligand and metal synthesis. The complexes thus prepared from these ligands can serve as useful catalyst for asymmetric synthesis. For example, a vanadium complex prepared from such ligand is very useful asymmetric oxidation catalyst (equation 5.4).

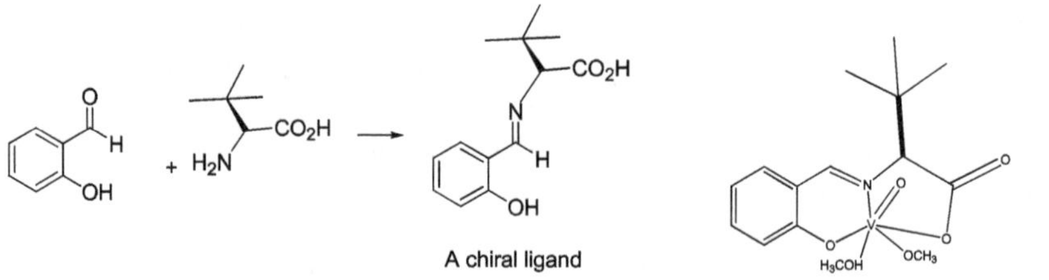

$$.....(5.4)$$

The development of reagent and identification of catalysts is important for organic synthesis. There is constant effort for development of such reagent. However, there are illustrative examples of reaction which are general in nature and can act as wide variety of substrates in very selective manner. One such example of catalysis is by cuprous bromide in the presence of tertiary butyl hydroperoxide. This catalytic system leads to various C-C bond forming reactions. Two such reactions are illustrated below (equation 5.5 and 5.6).

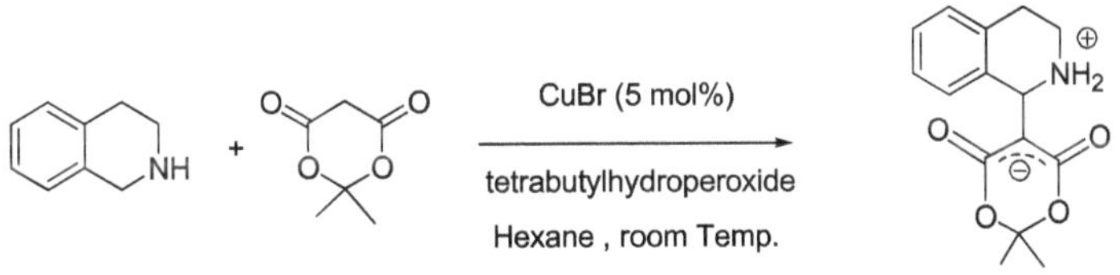

$$\ldots\ldots(5.5)$$

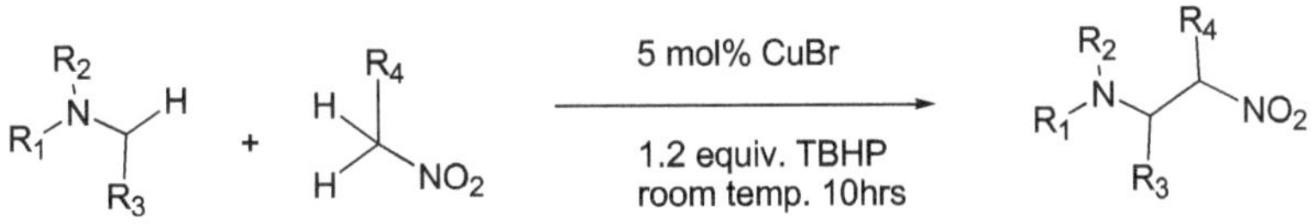

$$\ldots\ldots(5.6)$$

The mechanism of the reaction involves the steps shown in scheme 5.3.

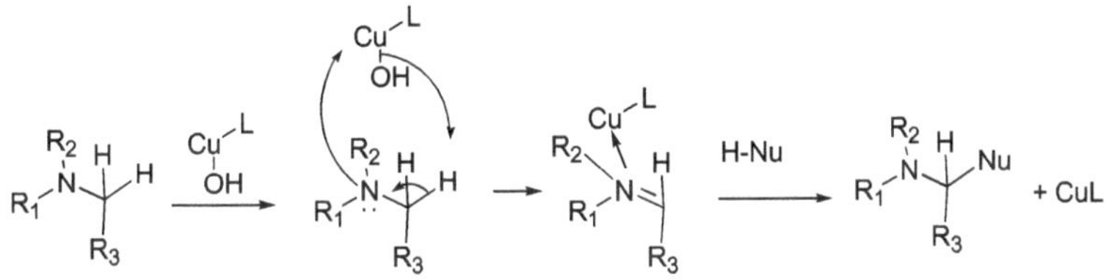

Scheme 5.3

It involves formation of a copper hydroxide species, which extracts a proton to form C=N bonded intermediate and such intermediates react with a nucleophile to give desired product (scheme 5.3).

Cyclodextrin

Cyclodextrins are oligosaccharides composed of 5 or more α-D-glucopyranoside units. Depending on the number of sugar units they are α-cyclodextrin comprising of six sugar rings, β-cyclodextrin with seven sugar rings, and γ-cyclodextrin with eight sugar rings. These molecules adopts cone like structure in which upper ring and lower rings have definite dimensions. The outer rim of the the α, β, and γ-cyclodextrins are 0.56 nm, 0.70 nm, 0.88 nm respectively. They are like bottomless bowl-shaped molecules with depth of 0.8 nm whose walls are stiffened by hydrogen bonding around the outer rim. The hydrogen bond strengths in different cyclodextrins are of the order α-cyclodextrin < β-cyclodextrin < γ-cyclodextrin (Fig. 5.9).

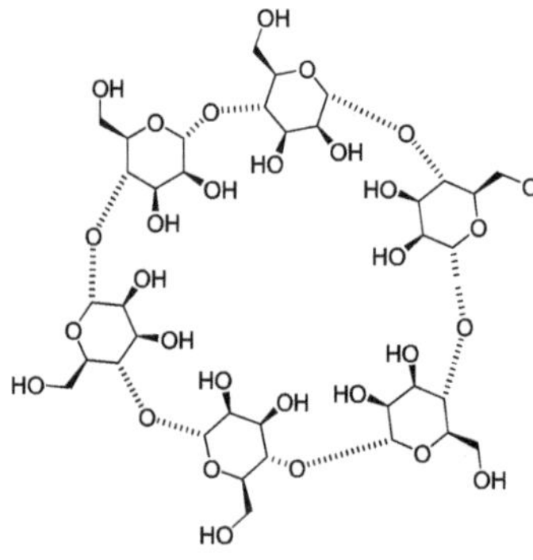

α-cyclodextrin

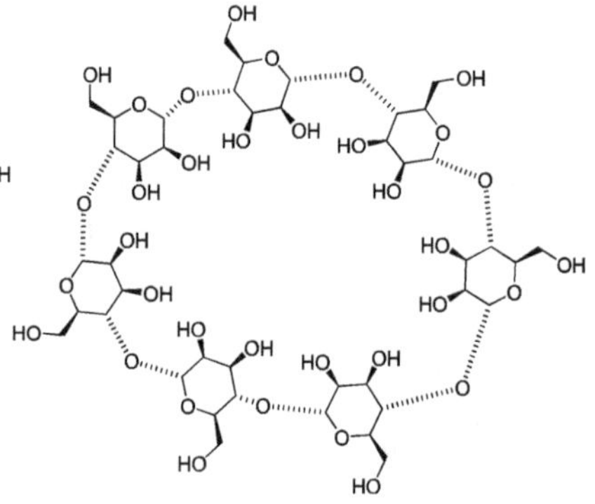

β-cyclodextrin

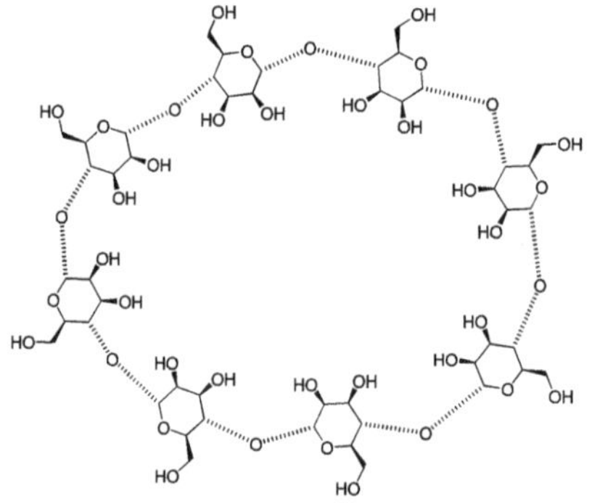

γ-cyclodextrin

Fig. 5.9 Different forms of cyclodextrins

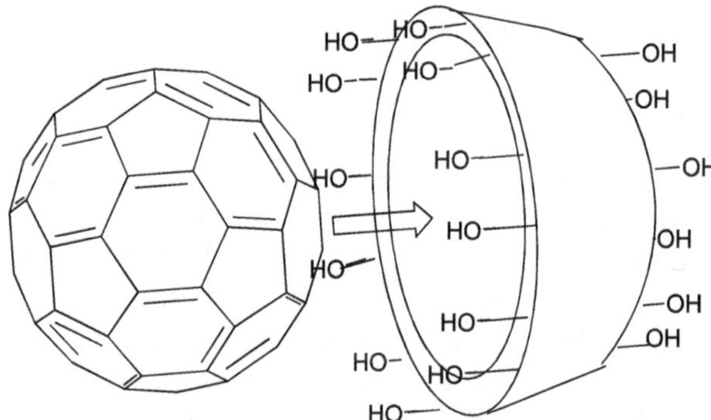

Fig. 5.10 Encapsulation of C_{60}

Due to their definite void sizes they form selective inclusion complexes and serves as template for selective organic transformations. The γ-cyclodextrin can selectively fit and bind to C_{60}-fullerene molecule as shown below. This binding can be simply achieved by mixing alcoholic solution of the γ-cyclodextrin with C_{60} (Fig. 5.10).

Nano particles

Nanomaterials find importance in electronic industry, catalysts, biomaterials etc. Among the various methods, one of the convenient methods is to prepare nano particles through solution chemistry. In these cases the metal ions at higher oxidation state are reduced to metals or corresponding oxides, sulphides etc through chemical reactions and precipitated from solution. These leads to particle size varying from nanometer to micron sizes. The colloidal materials have different UV-visible characteristic from their metallic or other form in bulk. The UV-visible study thus allows their formation, however exact size of the particles are determined by electron microscopy.

The metallic particles of nm sizes are generally not stable and they are stabilsed by adding stabilizer either in the form of surfactant or polymers such as sodium dodecyl sulphate, poly vinylalcohol etc. These are known as capping reagents. Their presence can be observed by recording IR spectra of the nano-particles.

Copper nano-particles can be prepared by adding sodium borohydride to a solution of copper(II) sulphate. The metallic copper is formed and they easily agglomerate; so to the solution sodium dodecyl sulphate is to be added to keep the entity of nano-particles intact. Dark brown colored copper nano-particles have plasmon transition at 580-600 nm.

$$CuSO_4.\,5H_2O \;+\; NaBH_4 \longrightarrow Cu \;+\; Na_2SO_4 \;+ H_3BO_3$$

$$.....(5.7)$$

The potassium permanganate in basic medium forms manganese dioxide. The advantage of this reaction can be taken to prepare manganese dioxide nano-particles. However, the particles formed immediately agglomerates to give micron size particles. The particle size can be improved by adding a surfactant cetyltrimethylammonium bromide to the solution (Fig. 5.11 and 5.12)

Scanning electron micrograph (SEM) of manganese dioxide prepared by reaction of sodium hydroxide with potassium permanganate under two different conditions are shown in Fig 5.11 and Fig 5.12.

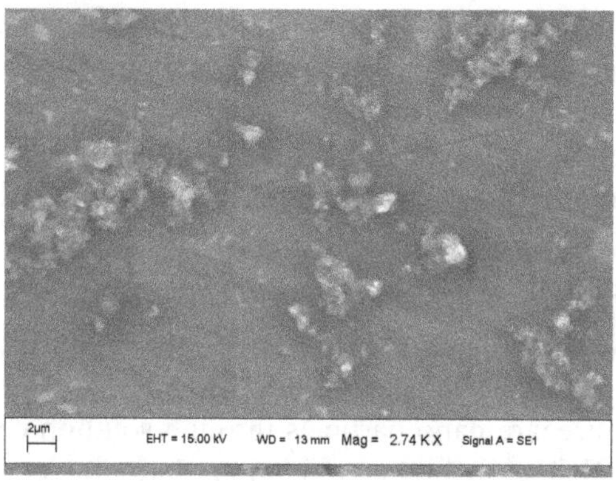

Fig. 5.11 SEM of MnO$_2$ from Sodium hydroxide with KMnO$_4$

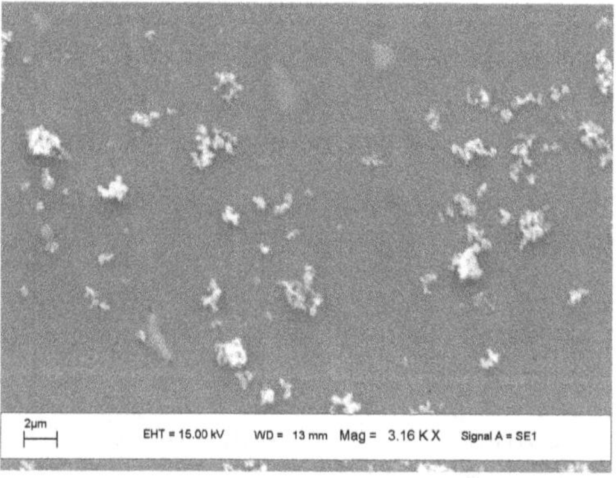

Fig. 5.12 SEM of MnO$_2$ from Sodium hydroxide with KMnO$_4$ in presence of cetyltrimethylammonium bromide

Preparation of manganese dioxide nano-particles

Prepare a solution of manganese(II) sulphate monohydrate (1.69 g, 10 mmol) in water (3 ml) and another solution of sodium hydroxide (2.5 mol) in 5 ml water. Add simultaneously these two solutions to a hot solution of potassium permanganate (1.9 g, 1.0 mmol) in water (15 ml). Stir the solution for 1 hr and isolate the brown precipitate of manganese dioxide by filtration. Dry the product in a dessicator under vacuum and grind finely. Repeat the experiment by adding cetyltrimethylammonium bromide (0.5 g) to manganese solution (in step 1). When scanning electron micrograph of these particles are compared it will be observed that in the former case micron size particles are largely formed whereas in the later case the nano-meter size particles are formed.

Synthesis of Fe_3O_4 nanoparticles

Reaction can be represented as follows:

$$Fe(acac)_3 + ROH + RCOOH + RNH_2 + Ph_2O \xrightarrow{\text{heat}}$$

Fe_3O_4 nanoparticle 4 nm

$$.....(5.8)$$

Take *tris*-acetylacetonato iron (III) (2 mmol) in diphenylether (20 ml). To this solution add 1, 2-hexadecanediol (10 mmol), oleic acid (6 mmol) and octylamine (6 mmol) under nitrogen atmosphere. Heat the solution to reflux for 30 minutes. Cool the reaction mixture to room temperature, a dark brown solid will be obtained. Wash this solid with ethanol (10 ml) under air; a dark brown precipitate will be obtained. Redissolve the precipitate in hexane in the presence of oelic acid and octylamine and re-precipitate with ethanol to obtain Fe_3O_4 nano-particles.

Polyaniline

The use of conducting polymers in electronic application has received much attention in recent years. Polynainline falls under this category of polymers. Thus, utility of polyaniline for applications such as organic lightweight batteries, microelectronics, optical displays, antistatic coatings, and electromagnetic shielding materials are well studied. Polyaniline is C-N bonded polymer of aniline and has (-B-NH-B-NII-) and oxidized (-B-N=Q=N-) repeating units where B and N denote benzenoid and quinoid ring respectively. In a chain of polyaniline the ratios of amine to imine varies and depending on the backbone they are classified as leucoemeraldine, emeraldine and pernigraniline.

The extent of oxidation decides their electrical properties. Polyaniline can be synthesized by oxidizing the aniline monomer by various means. The degree of conjugation differs depending on the method of preparation. So the method of preparation affects the electrical and optical properties of polynailine. Doping of the polyaniline can also control the electrical and optical properties of polyaniline. In the case of polyaniline, they can be protonated with a protonic acid or can be doped with charge transfer with oxidizing agents. Thus, it is possible to synthesise doped polyaniline, by using organic sulphonic acids containing large organic groups as dopant acids; which are soluble in various organic solvents. Transparent, blends of polyanilines with conventional polymers like poly (methylmethacrylate) can be prepared for optoelectronic use.

These films are also used as sensors for air borne volatile organic compounds like alcohol, ether, esters etc. On exposure of the vapors of analyte change in the resistance is observed. Interestingly the conductance of these materials can be measured by making film of it as well as blend and mechanism of conductance under different stimuli can be studied easily. Semiconductor properties can be confirmed by recording the conductance of such materials vs temperature as well as applied field at various temperature. Constructing a two or three probe electrodes system can do this and putting the samples between the electrodes placed over a heating block.

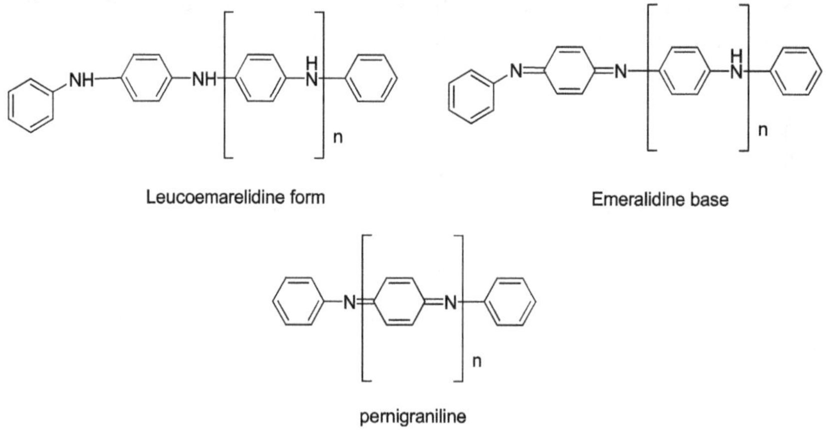

Leucoemarelidine form

Emeralidine base

pernigraniline

Fig. 5.13 Different forms of polyaniline

Preparation of polyaniline

Ployaniline can be prepared by oxidation of aniline by oxidant such as persulphate, hydrogen peroxide in the presence of a metal ion as catalyst. For example a ferrous ion catalysed oxidative polymerization reaction of aniline by hydrogen peroxide as shown in scheme 5.4.

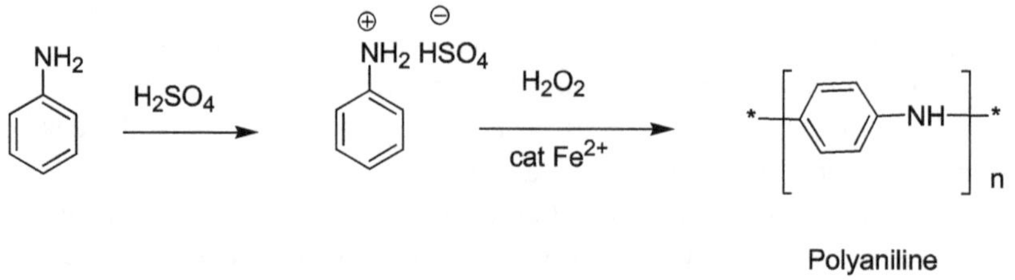

Polyaniline

Scheme 5.4 Formation of polyaniline

Procedure

Take aniline (1.4 mol) in acetonitrile (80 ml), add sulphuric acid solution (12 mmol, in 80 ml water) dropwise to this. After completion of addition sulphuric acid, add powdered ferrous sulphate heptahydrate (0.036 mmol), followed by hydrogenperoxide (4.5 mmol). Stir at room temperature for 6 hrs. A greenish precipitate of polyaniline will be formed on concentration of the solution. Filter and wash the polyaniline with water and dry.

Thin films by spin coating

Ordinary light is made up of seven colors, which has different wavelengths. When one of these colors is removed by interference, the complementary color will be seen. For example, when blue light is subtracted from white light, we observe yellow color. The surface of soap bubble or foam glistens with the complementary colors produced by interference phenomenon. The surface of soap bubble is made of thin film. Thin films are also useful for studying material properties such as conductivity, optoelectronic properties, and surface properties. However, to make a uniform film some techniques are needed. There are many methods, based on evaporation techniques or by condensation methods. Among them the spin coating is one conventional technique. In this technique a spin coater is used to prepare thin films of polymeric materials on glass-slides. Place a glass slide inside the spin coater and rotate it to its maximum speed. Put a drop of polymer solution (50 µl) at a time on the rotating slide and rotate the slide for 5 min. Remove the slide from the spin coater and check the color of the film in reflected light. A film that has a thickness of 400-700 nm will be colored in reflected light. Dry the film in a dessicator under vacuum.

To determine the thickness of a film following principle is adopted. Both the top and bottom surfaces of a thin film reflect light. The amount of reflected light is dependent on the sum of these two reflections. Depending upon the phase of radiation used, these two reflections may add together in a constructive or destructive manner. Thus, when a beam of light passes through a thin semitransparent film, it causes interference in the light wave as shown in the Fig. 5.14. This happens due the partial reflections from the surface. If the film thickness is of the order of 400-700 nm the interference fringes are observed (Fig. 5.14). Thus, the color of a thin film coated on a glass surface depends on the film thickness. The magnitude of absorbance by the film is of the order of wavelength of visible light.

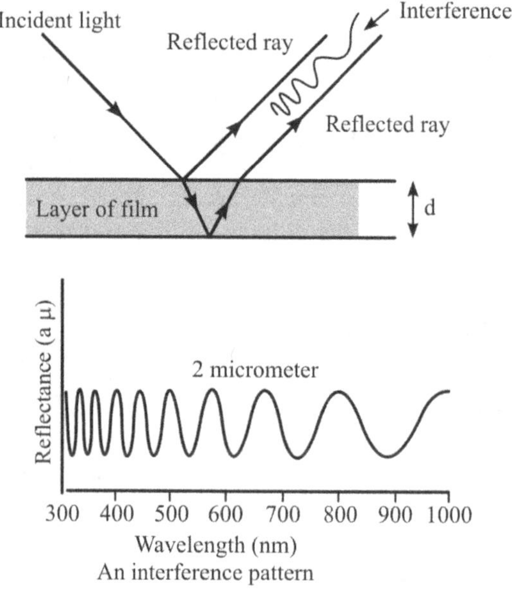

Fig. 5.14 Interference of light by a film

In the UV-visible spectra of the film a set of different maximum and minimum absorbance will appear. These maximum and minimum absorbance and separation between two maximum and minimum will be decided by the thickness of the film. The thickness of the polymer film can be estimated from the equation 5.9.

$$v = \left[\frac{n(\lambda_1)}{\lambda_1} - \frac{n(\lambda_2)}{\lambda_2} \right] / N_{cyc} \qquad \qquad .....(5.9)$$

$$d = (2v)^{-1}$$

where, $n(\lambda_1)$ and $n(\lambda_2)$ are the refractive indices of the thin film at wavelengths λ_1 and λ_2. N_{cyc} is the number of cycles in the interference fringes and d is the thickness of the film.

Making Films

Place one glass slide inside the compartment of the spin coater. Turn on the power supply and allow the fan to reach its maximum speed. Drop 50 μl (50-70 g/l) of the solution of sustrates whose film is to be prepared (for example polystyrene in acetonitrile) at a time onto the rotating slide and allow to rotate for 5 mins. Turn off the power supply and let the fan come to complete halt. Carefully remove the slide without touching the film. Judge the quality of the film by checking smoothness (eye estimate) and color of the film in reflected light. A film that has a thickness between 400 nm and 700 nm (roughly) will appear colored in reflected light. The color may be weak and experience is needed to detect the color or set of colors that appear on the film. Dry the films under vacuum for an hour and can be ready for further spectroscopic analysis.

High temperature Superconductor

Certain materials at very low temperature shows no electrical resistance and this phenomenon is called superconductivity. Tin, aluminium and various alloys shows superconductivity. However, most of these materials show superconductivity at such a low temperature that they are of very less use. Thus, high temperature superconductors are sought for. Lanthanum based cuprate materials shows high temperature superconductivity (about 90 K). Recently iron based superconductors namely lanthanum oxygen fluorine iron arsenide ($LaO_{1-x}F_xFeAs$) is observed to show superconductivity at about 26 K.

Preparation of a superconductor

The overall reaction for preparation of a yttrium barium cuprate super conductor is (equation 5.10)

$$Y_2O_3 + 2\ BaO_2 + 3\ CuO \ \text{------>}\ YBa_2Cu_3\,O_7 \quad .....(5.10)$$

Take yttrium hydroxide (0.60 g) and transfer them in a small beaker (make sure the beaker is dry). Weigh separately equivalent amounts of barium peroxide and cupric oxide transferring and place them in the beaker. Mix the three materials in a mortar. Make a pellet of the mixture and place in an alundum (a form of Al_2O_3) boat and heat it in a furnace. The furnace is heated to 930 °C over a period of about 8-12 hours, held at 930 °C for 12-16 hrs, cool the mixture to 500 °C and hold there for 12-16 hours. Finally turn off the furnace and cool the mixture to attain room temperature. Store the cooled pellets in a dessicator. The finished pellets are dark gray to black in color.

Meissner effect

Using the plastic forceps, place a superconductor pellet in a cup containing liquid nitrogen. When the pellet cools below the critical temperature, levitate a magnet above the superconductor. Touch the pellet gently with a forcep and the pellet starts to spin.

Organic superconductors

Bis(ethylenedithio)tetrathiafulvalene derivatives are organic high temperature superconductors. A synthetic procedure for preparation of precursor of organic super conductor is shown in scheme 5.5.

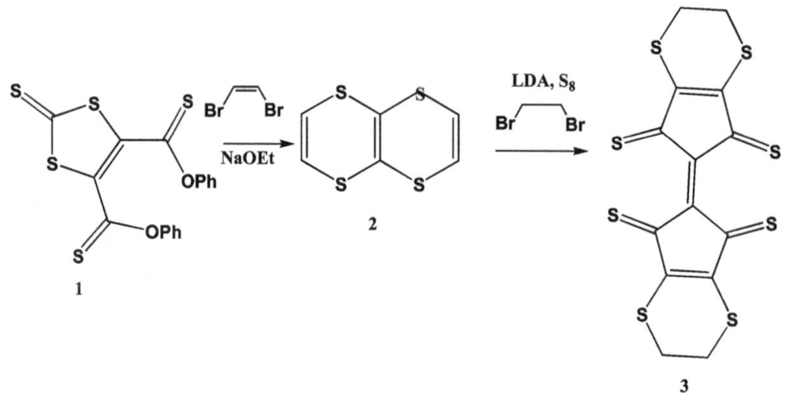

Scheme 5.5

Synthesis of 1,4,5, 8-tetrathianaphthalene (2)

Take fresh sodium (4.6 g, 200 mmol) in a 3-necked 1000 ml round bottom flask equiped with a stir bar, connect two 125 ml pressure equalizing funnels and a reflux condenser and make a nitrogen atmosphere. Add ethanol (75 ml). Once all of the sodium had reacted, add 250 ml of THF to the reaction maxture. Fill one dropping funnel with a solution of 4,5-bis(benzoylthio)-1,3-dithiole-Z-thione **(1)** (8.12 g, 20 mmol) dissolved in tetrahydrofuran (125 ml) and take another solution of dichloroethylene (3.2 ml, 42 mmol) dissolved in of THF (125 ml)in other dropping funnel. Add the two solutions dropwise simultaneously, over a 4 hr period under gentle reflux with stirring. Reflux the mixture overnight. Cool the solution and add 200 ml of water, dissolve the precipitate and the solution turns purple color. Remove the THF to obtain a light brown-yellow solid. Collect the solid by vacuum filtration. Recrystallize the product from cyclohexane/hexane (5:3) to give orange/yellow crystals.

m.p 125-127 °C. ^{1}HNMR (CDCl$_3$) 6.45 (s, 4H).

Synthesis of *bis*(ethylenedithio)tetrathiafulvalene (3)

Take BuLi (12.4 ml of a 1.6M solution in hexane, 19.84 mmol) with a solution of diisopropylamine (2.8 ml, 20 mmol) in dry THF (100 ml) in a round bottom flask (250 ml) and place at -78'C (dry ice/isopropanol) under N_2. Stir the solution for 1 hr. Add 4,5-bis(benzoylthio)-1,3-dithiole-Z-thione (**1**) (0.4 g, 1.96 mmol) in dry THF (30 ml) over a 30 min period Stir the solution at -78 °C for 3 hrs, add sulfur (0.26g, 8.1 mmol) in one portion. Stir the resultant brown solution was stirred at -78 °C for 1 hr and allow the reaction mixture to warm to 24 °C and leave overnight. After 24 hrs, add hexamethylphosphoramide (10 ml) into the flask at 24 °C and after one hour, add 1,2-dibromoethane (18 ml, 209 mmol) over a period of 2 hrs. After stirring for one day, extract the solution by soxhlet with carbon disulfide (100 ml) to obtain an orange solution. This on evaporation gives **3** as a bright red solid. Recrystallise the product from thiophene.^{1}HNMR (CDCl$_3$): 3.28 (s, 8H, CH$_2$)

Molecular magnets

Magnetic materials are important from inorganic material design. Various multimetal complexes serve as magnetic materials. Polyhedrons like structures of chromium complexes can be constructed by making mixed valent oxalate complexes of the type $\{Bu_4N\,[MCr(ox)_3]\}_n$ where M is a divalent cation such as Mn^{2+}, Fe^{2+}, Co^{2+}, Ni^{2+}, Cu^{2+}, Zn^{2+} etc. These compounds are easily prepared by reacting potassium *tris*-oxalatochromate(III) trihydrate with salts of other divalent metal ions.

$$K_3[Cr(ox)_3] + Mn(NO_3)_2 + (nBu)_4NBr \rightarrow [\{(nBu)_4N\}\{MnCr(ox)_3\}]_n + KNO + kBr$$

Where ox = oxalate dianion   .....(5.11)

Let us compare the *tris*-oxalatochromate(III) trihydrate with the following structure:

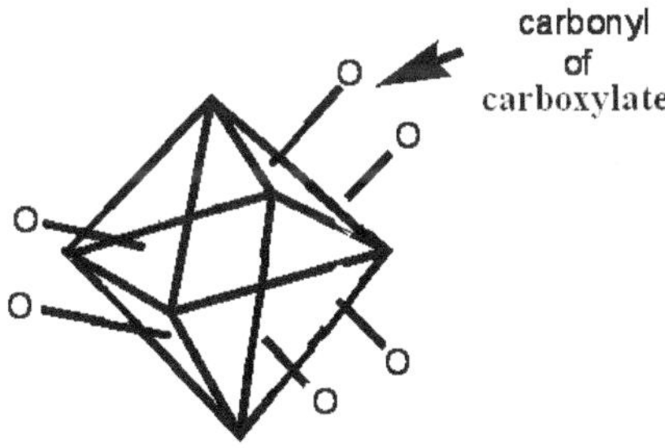

Fig. 5.15 Building block of *tris*-oxalatochromate(III) anion

In such a situation different polyhedron can be combined by using a divalent cation to generate 3-dimensional networks as follows (Fig. 5.16).

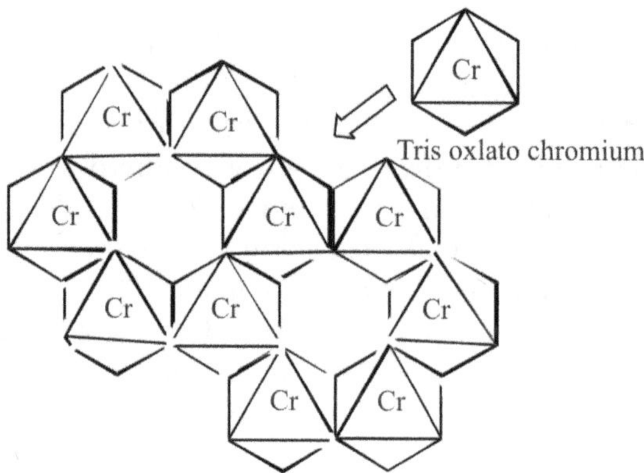

Fig. 5.16 Structure of a molecular magnet

These structures will make different orderly arrangngements which will be reflected in obtaining different magnetic properties depending on the metal ion used to share the octahedrons.

Preparation of [{(nBu)₄N} {MnCr(ox)₃}]ₙ

Take potassium *tris*oxalatochromate(III) trihydrate (2 mmol) in water (10 ml). To this add a solution of manganese(II) nitrate tetrahydrate (2 mmol) taken in water (10 ml). Stir the mixture and add *tetra*-n-butylammonium bromide (2.2 mmol) and this results in precipitation of the desired compound. A green crystalline compound will be obtained. The methanol solution of the compound absorbs at 692 nm, 574 nm and 424 nm.

Shape memory alloy

An alloy that can memorise its shape and can be returned to its original shape after it is being deformed is called a shape memory alloy. Shape memory alloys are applied in the field of medical and aerospace industries. They are generally made by casting, under vacuum arc melting or by induction melting. Some examples of shape memory alloys are nickel-titanium (Nitinol), gold-cadmium, brass etc. Shape memory effect can be represented by the following diagram by plotting deformation vs temperature (Fig. 5.17).

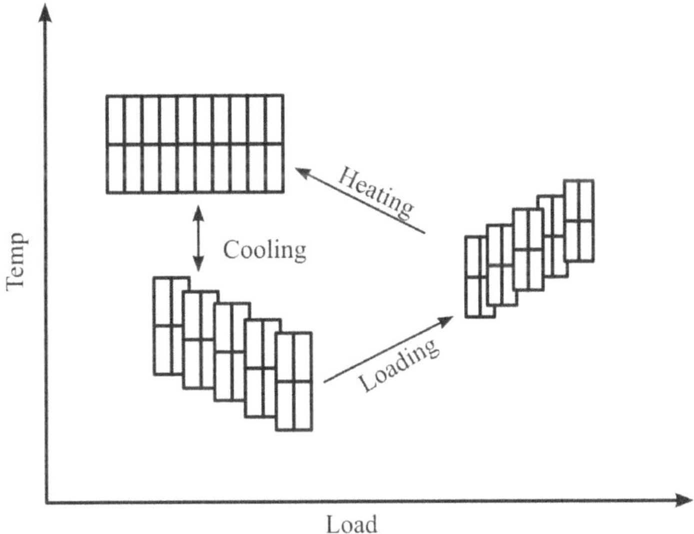

Fig. 5.17 Plot of deformation vs temperature of a shape memory alloy

When a piece of shape memory alloy is cooled alloy may be converted to form; which can be easily deformed. After deforming it on heating the original shape may be restored.

Sugar sensor

A sugar sensor can be devised from the following principles. Sugar can strongly bind with phenylboronic acid. Phenylboronic acid also binds to Alizarin red S. Upon binding of alizarin red with phenylboronic acid the salt formed is highly fluorescent on excitation at 490 nm it emits at 580 nm. However, on addition of diol or sugar this compound loses its fluorescence as it gives back the parent compound (scheme 5.6)

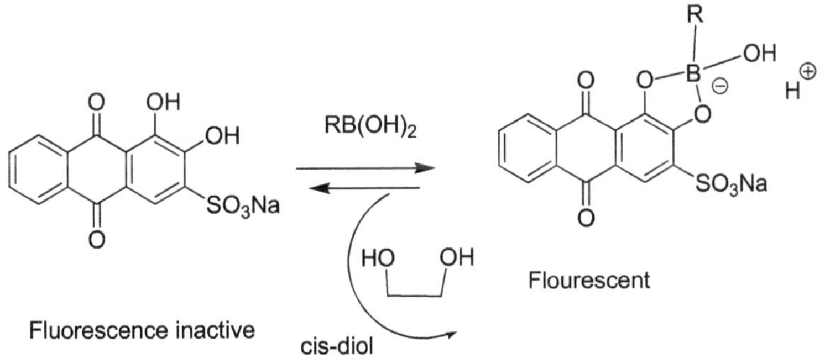

Scheme 5.6 A sugar sensor

Thus, when sugar is added to a preformed salt of alizarin with phenylboronic acid the fluorescence gets quenched. This system is very sensitive process and is used for detection of fructose and the system acts as a sensor.

Chiral derivatising agents

A chiral derivatizing agent is a molecule that can be easily derivatised with a mixture of enatiomers that gets converted to diasteromers. The diasteriomers thus formed can be analysed by spectroscopic or chromatographic technique to ascertain the enantiomers present in the mixture.

$$.....(5.12)$$

Enantiomerically pure form (S)-(+)-N-acetylphenyl glycineboronic acid reacts with chiral diols and gives quantitatively cyclic boronic esters. These esters show a remarkably high diastereo-differentiation of proton NMR signals and thus they are used to determine enatiomeric excess of chiral diols.

Perfumes

The chemist working on perfumes can be considered artists for making fragrances. This needs extra skill of combining creativity with the knowledge found in areas such as analytical, synthetic, and physical chemistry to aestheticism. In earlier days the use of aromas was only for worship practices and now it has become a part of day to day use. Some of natural components that contribute significantly to production of different smells are listed in Fig 5.18.

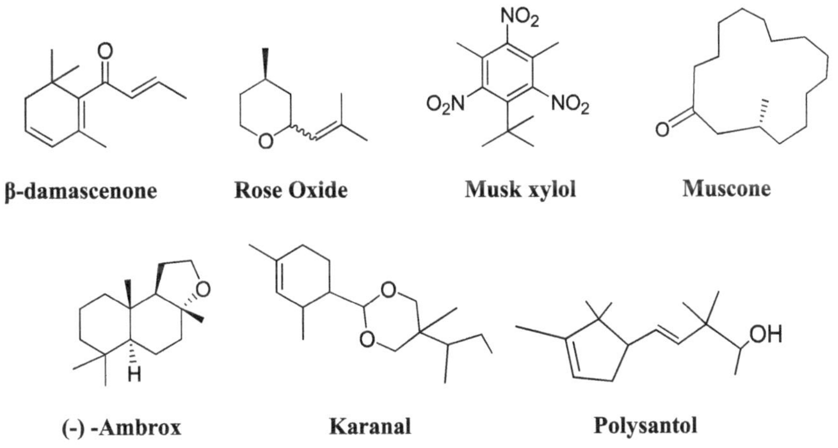

β-damascenone　　　　**Rose Oxide**　　　　**Musk xylol**　　　　**Muscone**

(-) -Ambrox　　　　**Karanal**　　　　**Polysantol**

Fig 5.18 Different compounds used as fragrance

These components are mixed with different alcohols or mixed solvents. In many cases additives are added to increase the different needs, which include smell affinity and intensity of smell.

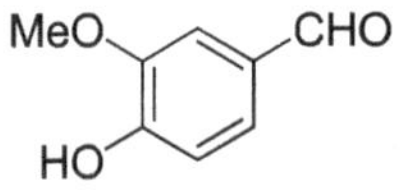

Fig. 5.19 Structure of vanillin

One such commonly used additive is vanillin (fig 5.19). The variation of ether substrate instead of methyl ether in such cases changes the smell. Coloring materials are added to increase the aestheticism.

Non linear optical materials

Nonlinear optics deals with the interaction of electromagnetic radiation with a substrate in which the responds of the radiation takes place in a nonlinear manner to the incident radiation. The nonlinear response causes change in intensity of the propagation of the radiation or in the creation of new radiation fields with new frequencies or changes the directions of propagation. Such effect may occur in all the states of matter. The far infrared to the vacuum ultraviolet region is generally chosen for such study. One or combination of two or more radiations may generate the non-linear optical effects. There are many aspects that are studied in non-linear optics. But to put in the most simplified aspects of non linear optics to a chemist is the term called second harmonic generation is generally emphasized. In sense it may be explained as an effect in which a molecule absorbs two photons the energy of they may double by this interaction, in that case it is called intensity doubling. If the intensity is not double but there is an enhancement in intensity, the process is said to occur through intensity mixing. The molecules that have a donor and acceptor at two ends are molecules of interest for such properties. Some examples are given in Fig. 5.20.

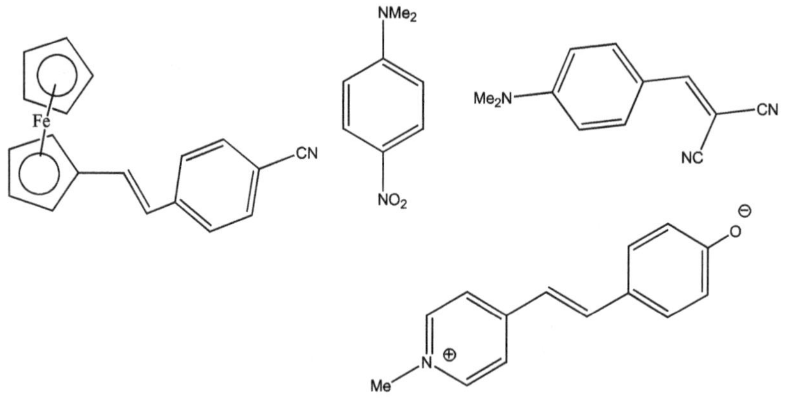

Fig. 5. 20 Some examples of molecules having second order nonlinear susceptibility

Molecules possessing large permanent dipole are although of interest for nonlinear optical study but these crystallize in centrosymmetric space groups making them less suitable for such studies.

Concluding remarks

The area of materials design has no bound of limit. Whether it is a polymer or supramolecular assembly or single molecule, any of them showing specific properties are of interest to chemist. The gels, liquid crystals, molecular devices, optoelectronic materials, fuel cells etc will be generated as new materials. The experimental tools and new experimental means to analyze the new properties will continue to be a central point for study to a chemist. Hence, there is a need of emphasis on analytical chemistry without putting a barrier of conventional branches in chemistry.

Index

M

N

O

P